THE
ANTHROPOCENE

THE
ANTHROPOCENE

THE HUMAN ERA
AND HOW IT SHAPES OUR PLANET

CHRISTIAN SCHWÄGERL

Foreword by PAUL J. CRUTZEN

Translated from German by Lucy Renner Jones

SYNERGETIC PRESS
Santa Fe & London

Published by Synergetic Press
1 Bluebird Court, Santa Fe, NM 87508
24 Old Gloucester St., London WC1N 3AL

Original title: Menschenzeit. Zerstören oder gestalten? Die entscheidende Epoche unseres Planeten, by Christian Schwägerl, © 2011 by Riemann Verlag, a division of Verlagsgruppe Random House GmbH, Munich, Germany

Library of Congress Cataloging-in-Publication Data

Schwägerl, Christian.
  [Menschenzeit. English]
  The anthropocene : the human era and how it shapes our planet / Christian Schwägerl.
    pages cm
  ISBN 978-0-907791-54-6 (hardcover) -- ISBN 0-907791-54-9 (hardcover) -- ISBN 978-0-907791-55-3 (pbk.) -- ISBN 0-907791-55-7 (pbk.) 1. Environmental responsibility. 2. Environmental ethics. 3. Nature--Effect of human beings on. I. Title.
  GE195.7.S3413 2014
  304.2--dc23
                              2014034709

Book and cover design by Barbara Haines
Editors of the English language edition, Linda Sperling with Hugh Elliot
Translated from German by Lucy Renner Jones
Printed by Global PSD, USA on 55# Glatfelter White D-37 stock
Typefaces: Minion with Scala Sans display

CONTENTS

FOREWORD

THIS BOOK IS A STRIKING MANIFESTATION of the power and potential in the idea of the Anthropocene. A skillful narrator with many years of journalistic experience, author Christian Schwägerl describes how a single species, our own, is irreversibly transforming the Earth's biological, geological and chemical processes, and thus affecting our very existence. Two hundred years of industrialization bear testimony to humanity's power of innovation and creativity but also prove our more perilous powers of degradation and destruction. For the first time in Earth's history, its future is being determined by both the conscious and unconscious actions of Homo sapiens. Schwägerl's book is a rallying cry for us to recognize our opportunity to build a long-lasting, viable, creative and freedom-loving human civilization. This book is like a navigation system for the new world of the Anthropocene that lies before us.

Paul J. Crutzen, PhD

Dr. Paul J. Crutzen, born 1933, is an atmospheric chemist who won the Nobel Prize for Chemistry (with Mario J. Molina and Frank Sherwood) in 1995, for his pioneering research into ozone layer depletion caused by chlorofluoro-carbons (CFCs). From 1977 to 1980, Dr. Crutzen was the director of research at the National Center of Atmospheric Research (NCAR) in Boulder, Colorado and from 1980 to 2000, director of Atmospheric Chemistry at the Max Planck Institute for Chemistry in Mainz, Germany. He has undertaken research at numerous other institutions, such as the Scripps Institution of Oceanography and the Georgia Institute of Technology. Crutzen is a long-standing member of various scientific academies, including the National Academy of Sciences and the American Academy of Arts and Sciences.

PREFACE

THINK I CAME ACROSS THE WORD Anthropocene a few times without it making an impression. Then, one day, it really struck me. I was at a lunch with my friend Matthias Landwehr, in 2008, and he told me that this subject was something I should investigate. I agreed to do so. When I got back to my desk, I had an epiphany and my twenty-five years of reporting on environmental and science issues and forty years of love for everything natural suddenly appeared in a new light, as if I had been touched by a magician's wand.

One man, Paul J. Crutzen, had defined the relationship between humans and planet Earth, in such a powerful way, that it was hugely inspiring. Crutzen had melded humans and nature (two entities that I had previously thought of as separate, opposing forces), into a whole new science-driven idea. It described a connection that reaches back into the past and far into the future. After seeing, at first hand rainforests burning, land made toxic from mining, and species on the brink of extinction, this idea gave me hope that our ever evolving human consciousness might be about to enter a new phase.

So, this is how I came to write *Menschenzeit* (The Age of Humans), the German language precursor of what you hold in your hands. *Menschenzeit* was launched at the Berlin Museum for Natural History, in September 2010. Achim Steiner, the head of the United Nations Environment Programme, came to Berlin for the occasion and gave a wonderful speech about freedom and responsibility. The book launch also resulted in the beginning of a productive friendship with Reinhold Leinfelder, then director general of the Berlin Natural History Museum. Together, we approached the Haus der Kulturen der Welt, (House of World Cultures, abbreviated as HKW), a state-funded cultural center situated right next to the Federal Chancellery, Max Planck Research Society, and the Deutsches Museum in Munich, one of the world's leading technology museums, and proposed the idea of Anthropocene. We suggested this was a subject both institutions might want to pursue. What followed then was three years of productive collaboration and inspirational events at the HKW. "The Anthropocene Project" was funded directly by the German parliament. The first large-scale exhibition, "Welcome to the Anthropocene—The Earth in our Hands" is scheduled to be held at the Deutsches Museum, from December 2014 through January 2016.

I was able to convince Paul Crutzen to serve as an honorary patron of both projects and I'm very grateful that I have been able to discuss the Anthropocene idea with him so often in recent years.

When I was researching and writing the German edition of this book, the idea of the Anthropocene was rarely mentioned in the media. It might well have become an intellectual cul-de-sac. But, because of its inspirational quality and the efforts of many luminaries, including Jan Zalasiewicz, Reinhold Leinfelder, Andrew Revkin, Will Steffen, Libby Robins, Jürgen Renn, Klaus Töpfer, Bernd Scherer and Helmuth Trischler, the idea has now gained traction. It is now being discussed around the world as a new perspective on how humans and animals, plants and stones, the oceans and all other components of the earth interact. As an author, it is a gratifying experience to see that *Menschenzeit* really did focus attention on the Anthropocene idea.

My work on "The Anthropocene Project" and preparations for the "Welcome to the Anthropocene" exhibition have allowed me to keep thinking about the idea of a man-made geological epoch and to join in many debates, see artists working with the idea, and to meet very interesting people. This has helped me to further refine my personal perspectives on the many questions stimulated by the Anthropocene idea. The result of all this is an updated English language edition of *Menschenzeit*, published by Deborah Parrish Snyder and her wonderful team at Synergetic Press.

There are three changes in my own perspective that have occurred since the German edition was published that I would like to highlight.

Initially, I considered it strange that the Anthropocene idea was associated with geology as opposed to biology. I thought that in order for it to be meaningful, it had to have a connection with the living world. What I have come to understand is that geology offers that connection, on a grander scale. By being geological, the Anthropocene opens a doorway between supposedly dead matter and living matter. It tells us that humanity and the technosphere it has produced are now participating in the largest and most long-term of planetary cycles, with conscious thought thrown in! In the words of political scientist Jane Bennett: "If matter itself is lively, then not only is the difference between subjects and objects minimized, but the status of the shared materiality of all things is elevated." After so many decades of a consumerist materialism that "treats the planet like a zombie" (Giulio Tononi), we now have a chance to develop a "vital materialism" that honors life in all its forms, including those made of stone. Thus, the Anthropocene idea becomes the opposite of anthropocentrism. You will find far more references to this impression of the Anthropocene in the current edition than there were in the original German language one.

I also once thought that a flaw in the Anthropocene idea was that it did not immediately state what was good or what was bad. Stripped bare, it's a scientific hypothesis about the geophysical state of the planet and does not take into account ethical or moral or spiritual values nor discuss the suffering of humans or other species. Since then, I have experienced personally how contemplation of the Anthropocene idea triggers strong, ethics-driven reactions and a strong impulse of caring. Sustainability ideas come prepackaged with a set of imperatives. The Anthropocene idea works differently, but in a complementary way. It exposes us all and asks for responsibility. It invites commitment and responsible behavior instead of demanding it. There is a possibility that this idea might be abused in order to advocate human entitlement and insist upon simple techno-fixes. However, I'm confident that this line of thinking will not prevail.

Now I am speaking, even more directly, against misanthropy and "doomsday-ism." I disagree with those who say the words "good" and "Anthropocene" should probably not be used in the same sentence. Why not? Are our children and their descendants already doomed to live through millennia of ecological hardship? Even worse, are humans the problem and should they therefore vanish from the planet? No! I am the last person to downplay the severe problems created by our civilization, many of which you will read about in this book. But despising the human species and waiting for the end of the world, as we know it, is not the answer.

The Anthropocene is more than the sum of the parts of environmental havoc. It can be the arena in which humanity decides to wisely integrate into the planet's workings, enriching itself by its actions as a result. Smart cities, cultivated life-forms and landscapes with a human-induced biodiversity, are examples of how we can create a positive geological record. Human creativity, community spirit and conscious thought can lead to changes that might make our species look back at current behavior as sheer ecological barbarism. This is the journey I invite you to take with me in this book: going from today's crises to an enlightened planet with beautiful human imprints.

Christian Schwägerl, Berlin, 2014

PROLOGUE **Writing in the Sky**

ON DECEMBER 3, 1933, in Amsterdam, Anna Crutzen, a woman in her early twenties, gave birth to a son. She had moved from the Ruhr region in Germany to the capital of the Netherlands, five years earlier, to earn her living as a housekeeper. She had met Josef Crutzen, a young man from Vaals, a small Dutch town on the German border, who was working as a waiter. They fell in love, married in 1932 and soon started a family.

They named their first child Paul Jozef. Nothing at that time indicated that this young boy would literally be responsible for saving the planet from an existential threat and would introduce a groundbreaking idea that would redefine humanity's place on Earth. Paul did not enjoy a private education like Alexander von Humboldt, nor did he have a botanically minded uncle, as did Charles Darwin.

He grew up in harsh conditions. His mother, who made many sacrifices to care for the family, worked as a steward in a hospital kitchen. His father was regularly unemployed and the family was very poor. In addition, the darkest period of the twentieth century had just begun. Only months before Paul's birth, Adolf Hitler, the new German Chancellor, had seized power in neighboring Germany. In 1939, just before Paul's sixth birthday, the German dictator ordered his army to invade Poland, starting World War II. Crutzen's childhood took place in the midst of war. The boy saw German troops march into Amsterdam and commandeer his school. He lived through the *Hongerwinter* (the famine of 1944–45) in which thousands of Dutch citizens died, including some of his friends. The sight of the Allied bombers that flew from England, across the Netherlands to bomb German cities caused him much distress. His mother's family lived across the border in the Ruhr, the industrial heartland of Germany, an area laid waste by daily and nightly Allied bombing.

In spite of the wartime conditions, his parents noticed that their son had a special talent for observation and a keen thirst for knowledge. He quickly learned German, French and Flemish, and even memorized dictionaries, just for fun. One bitterly cold winter night, his parents found him sitting, shivering in his pajamas by an open window, gazing up at the sky. Upon seeing snow for the first time, he didn't feel the cold. Paul often came up with unusual observations. When he first glimpsed a half-moon, he said it was "broken," and when he saw a man swimming in an Amsterdam canal he insisted for a long time that it was a head without a body. As a teenager, he not only played football but also began reading everything he could find concerning natural science.

After the war, Paul did not want to be a burden to his parents and realized that further education in natural science was beyond his means. There was just enough money for him to attend engineering school. He learned how to build bridges across the many canals in the Netherlands. Then, at the beginning of the 1950s, a life-changing incident occurred. As a child, Paul had always longed to see mountains. Holland is not exactly famous for its high peaks so he often fantasized that the cumulus clouds at which he liked to gaze were mountains.

Now that the war was over, it was possible once again to travel. Using his modest savings, Crutzen managed to get to Switzerland. Had he reached the summit of Mount Pilatus—a well-known mountain near the city of Lucerne—either ten minutes earlier or ten minutes later, he might have continued life as a bridge-builder but, just as he reached the summit, a young woman from Finland was starting her descent. She was working as an au pair and was spending her day off learning about her host country. Terttu Soininen walked past the young Dutchman just as he was about to take a photograph of the view and the two of them started talking.

———————

They married a few years later and moved to Sweden, to the small town of Gävle, to be nearer Terttu's family. Paul found a job as a construction engineer and began building houses instead of bridges. But his taste for knowledge, exploration and understanding had not diminished. The construction job only partly satisfied him. Thus, one day he glanced at a job ad in the newspaper: The Meteorological Institute at the University of Stockholm had an opening for a computer specialist. Admittedly, he hadn't the slightest experience in either meteorology or computer science, but something told him that he should apply.

The convoluted path leading Paul Crutzen to that café table in a provincial Swedish town, where a newspaper want ad lay before him, was to have momentous consequences—not only for himself but for all of humanity and for the course of Earth's history.[1]

Rare individuals sometimes change the course of human history on a large scale, in both positive and negative ways. We can think of examples as diverse as Alexander the Great, Jesus Christ, Julius Caesar, Christopher Columbus, and, more recently, Mikhail Gorbachev. But can ordinary human beings also alter the course of Earth's history?

———————

Yes, they can! A few years before Paul Crutzen was born, an American mechanical engineer and chemist named Thomas Midgley had unknowingly done just that. Midgley worked for the General Motors Chemical Company and had the task of developing new coolants for use in refrigerators and recently-invented air conditioners. Refrigerators had already been through their first design cycle. For all their benefits in helping to preserve food, refrigerator coolants were also volatile, poisonous and combustible. Midgley went in search of alternatives. He and his team produced synthetic compounds that didn't occur naturally. The substances mixed together in Midgley's laboratory included carbon, hydrogen, chlorine and fluorine, which, in turn, produced new compounds. One of these substances—chlorofluorocarbon, or CFC—proved to be ideal: it was odorless, non-toxic, highly stable and perfectly suitable for refrigeration. CFCs quickly entered the market under the trade name Freon® and were successfully used all over the world.

In the economic boom of post-WWII, millions of people, especially Americans and Europeans, were suddenly able to afford cars and televi-

———

1. Source: extensive interview with Paul J. Crutzen, 2013 and: http://www.nobelprize. org/nobel_prizes/chemistry/laureates/1995/crutzen-lecture.pdf.

sion sets, kitchen appliances, larger homes and long-distance travel. New supermarkets featured an abundance of foods as agriculture became more mechanized and productive. Chemical factories created new products that were designed to make life easier and more pleasant and that list included Freon. In the 1950s and 1960s, CFC use sharply increased because owning a refrigerator was now taken for granted. However, it wasn't long before discarded refrigerators dumped in landfills began leaking CFCs into the atmosphere. And because CFCs were odorless and colorless, no one noticed what was happening.

Unwittingly, Thomas Midgley's invention was causing immense damage to one of the systems most vital to protect life on Earth—a system that had taken hundreds of millions of years to evolve and that had made it possible for humanity to develop in the first place: the ozone layer. At a height of between six and thirty miles up in the Earth's atmosphere, the ozone layer intercepts most of the sun's harmful ultraviolet rays. Without the ozone layer, life on Earth, at least on land, would hardly be possible for higher diurnal life forms.

The environmental historian John McNeill later wrote that Midgley "had more impact on the atmosphere than any other single organism in Earth's history."[2] In the 1960s, no one knew about the dangers posed by chemicals mixing together, high above Earth's then three billion inhabitants. CFCs, nitrous-oxide, also known as "laughing gas," another ozone-depleting substance used in farming, and the burning of fossil fuels like coal and petroleum, were causing tremendous damage. Not even Paul Crutzen knew.

In 1958, our Dutch engineer applied for the position of computer specialist at the University of Stockholm's Meteorological Institute. He got the job because interviewers at the Institute believed that he would be a fast learner. This was the young man's entry into the world of scientific research, the career of his dreams. Not only was he a fast learner in the field of computer science, he also began to attend lectures in mathematics, statistics and meteorology. In 1963 he graduated and began a career as a scientist. Without deliberately planning it, he found himself in one of the hotspots of

2. John R. McNeill, *Something New under the Sun: An Environmental History of the Twentieth-Century World.* New York: Norton, 2001.

global environmental research. Among the young professors at Stockholm was Bert Bolin, who went on to co-found the Intergovernmental Panel on Climate Change (IPCC), which he then chaired, from 1988 to 1997.[3]

Crutzen picked an area of research that was relatively new: the chemical processes that take place in Earth's upper atmosphere. At first, he didn't realize how significant the gaps in then-current knowledge would prove to be. He was interested in the natural processes and how the protective layer of ozone in the atmosphere, constantly renews itself.

Then, his impact on Earth's future really began: he was one of the very first scientists to ask whether there are chemical processes that harm the ozone layer. Until then, this idea had seemed quite improbable. "The general feeling at the time was that 'nature is so big and mankind so small,'" says Crutzen today, in retrospect. "Nobody had thought that man-made substances could have a huge effect on stratospheric ozone."[4] Crutzen's initial research into how nitrous oxide, naturally released by soils, might damage the ozone layer, led him to a quite different discovery—that human activity was a threat to the ozone layer.[5]

Ever-increasing quantities of CFC molecules released into the atmosphere from landfills were not the only threat to the ozone layer. At around the same time, aviation engineers were developing large, high-tech nitrous oxide turbines. The United States, France, Great Britain and the Soviet Union had plans to build fleets of Supersonic Transport Aircraft (SSTs) to make it possible for civilians and the military to travel at supersonic speeds. Together with the American chemist Harold Johnston, Crutzen recognized the harm caused by nitrous oxide being released into the stratosphere so he used cool scientific logic to counter the dream that humans should be able to travel like gods to anywhere on Earth in a few hours. At the beginning of the 1970s he did some meticulous calculations to show that the additional nitrous oxide that would be emitted into the atmosphere from a fleet of 500 high-flying SSTs could cause "serious

3. I interviewed Bert Bolin in this function during the UN Climate Change Conference (COP1): Christian Schwägerl, "Umweltexperte: Kosten leider kein Thema in Berlin", *Süddeutsche Zeitung,* April 7, 1995.

4. From an extensive interview with Paul Crutzen in summer 2013.

5. Described in: Paul J. Crutzen, "Estimates of Possible Variations in Total Ozone Due to Natural Causes and Human Activities", *Ambio,* vol. 3, no. 6 (1974): 201–210.

decreases in the total atmospheric ozone layer and changes in the vertical distributions of ozone, at least in certain regions."[6]

The work of Paul Crutzen and his colleagues sent a new message to the world: humanity had become so powerful and dominant through science, technology and modern lifestyles that we could harm Earth's protective ozone layer.

Inspired by these warnings, other scientists began to look for additional chemicals that might be changing the ozone layer. Mario Molina and Sherwood Rowland made a discovery in 1974: they established that CFCs are particularly effective in destroying ozone molecules. When Crutzen heard about this groundbreaking work, he immediately contributed his research and calculations showing that the hypotheses of his American colleagues were correct and that in the foreseeable future, forty percent of the world's ozone layer might be depleted if the use of CFCs went unchecked. The consequences would be devastating; incidents of skin cancer and genetic mutation would multiply and some regions of the Earth might well become unfit for human life.

There were many objections to the researchers' theories, especially from the chemical industry that feared for its profits from the sale of CFCs and artificial fertilizers. The "hole" in the ozone layer could be a natural occurrence, critics argued. The harmfulness of CFCs was not proven. It would cause enormous damage to the economy to ban them, since there were no alternatives.

But in the mid-1980s, polar researchers led by Joe Farman returned from Antarctica with data showing that above a continent not populated by humans, the ozone layer was shrinking, especially during the southern hemisphere's spring and that hazardous ultraviolet rays were reaching the Earth's surface without hindrance. In 1985 Farman and his team published their results,[7] revealing why the thinning of the ozone layer was over Antarctica, of all places, and not over industrial areas: CFCs adhere particularly well to ice crystals.

6. Paul J. Crutzen, "My life with O_3, NO_x, and other YZO_xs". Stockholm: Almqvist & Wiksell International, 1995. See also: http://www.nobelprize.org/nobel_prizes/chemistry/laureates/1995/crutzen-lecture.pdf.

7. Joe Farman et al: "Large losses of total ozone layer in Antarctica reveal seasonal CIO_x/NO_x interaction", *Nature*, vol. 315.

The Antarctic explorers' findings shocked public opinion even more than Rachel Carson's *Silent Spring*. Before this, millions of people had associated their refrigerators and freezers with the idea of cold beer and instant pizzas as well as other convenience foods. They sprayed underarm deodorant in the mornings to smell nice at work in the office where, on hot days, they were grateful for air conditioning. Now, all of a sudden, these symbols of modern prosperity were seen in a completely different light. What had been thoughtlessly emptied into the atmosphere was suddenly coming back to haunt, in the form of a "hole" in the ozone layer. The individual actions of millions were a major hazard that threatened human life on Earth, and risked damaging the conditions needed for any terrestrial life.

No matter how aggressively the chemical industry opposed demands made by environmentalists and scientists like Crutzen, Molina and Rowland to ban CFCs, the case was won in 1987. The United Nations drew up the Montreal Protocol, the single most effective international environmental treaty to date, which called for a gradual phase-out of harmful CFCs, used primarily as coolants. In 1997, the Kyoto Protocol was ratified, which aimed to drastically cut carbon dioxide emissions and nitrous oxide (N_2O) emissions.

Since then, a slow but constant regeneration of the ozone layer over Antarctica has taken place. In 2012 and 2013, scientists at NASA and the Alfred Wegener Institute in Germany reported that the Antarctic ozone "hole" had become noticeably smaller, for the first time.[8] There is now a chance that the Antarctic ozone "hole" will disappear by 2050––so long as climate change and the extreme durability of CFC molecules do not thwart this.[9]

Currently, researchers are optimistic that the ozone layer is regenerating globally and will be permanently restored. What would have happened if Paul Crutzen had not survived the "hunger winter" in Amsterdam and no one had undertaken research like his?

8. See: http://www.nasa.gov/content/goddard/antarctic-ozone-hole-slightly-smaller-than-average-this-year/.

9. If increasing amounts of carbon dioxide in the lower atmosphere prevent the sun's rays from reaching Earth's surface, it will become colder where the ozone layer is. Such an effect led to the first formation of a large ozone hole over the North Pole in 2011, to the surprise of explorers. See: Gloria L. Manney et al., "Unprecedented Arctic ozone loss in 2011", *Nature*, vol. 478, (2011): 469–475.

What if he and his colleagues had not had the academic freedom and sufficient funding to explore the chemistry of the ozone layer without any specific aim?

What if Thomas Midgley had put much more aggressive and faster-acting bromine instead of fluorine into refrigerators and aerosol sprays, right from the beginning, well before reliable instruments for measuring the chemistry of the atmosphere existed?

What if explorers like Farman hadn't spent long nights and bitterly cold days in the Antarctic to make measurements that, at the time, did not interest anyone?

Questions like these concern Paul Crutzen, who says that he has often asked himself since then: "What other surprises may await us?"

The repair of the ozone layer in the twentieth century was dependent on many coincidences. Models show that the ozone layer could have completely disappeared by 2050 had CFC emissions persisted. When Crutzen received the Nobel Prize for Chemistry in 1995, together with Molina and Rowland—or perhaps we should say the "Nobel Prize for Salvaging the Ozone Layer"—he conveyed how utterly humiliating it would have been for humanity to have destroyed the atmospheric layer that protects life on Earth, by the expedience of using aerosol sprays and refrigerators, unaware of the damage being caused. "I can only conclude that mankind has been extremely lucky."

So, in the twentieth century, individuals were indeed making human history but also global history. On the one hand there was Thomas Midgley, the inventor of a substance that was endangering the hundred-million-year-old life-protecting atmospheric layer, and on the other, there were Crutzen, Johnston, Molina, Roland and others who, having recognized what was happening, demanded action. Events high up in the sky, were being determined not just by the interaction of molecules, temperature and pressure, but also by the work of chemists synthesizing new substances, and by the scientists who were investigating the effects of these substances. Human work manifest in the form of notes, index cards, laboratory diaries and scientific papers led the way to a new global reality.

For the first time in Earth's history, the results of human activity could be read, as if written high up in the sky.

Paul Crutzen was possessed by his discoveries about the ozone layer.

So, back at his desk at the Max Planck Institute, he set about making a list of the ways in which humans were transforming the planet. His list was long—and it grew longer. The more aware Crutzen became of everything that humanity was doing to the Earth, the more a new idea began to form in his mind. He realized that the prevailing view that mankind is miniscule whereas nature is limitless, and that humans only scratch the surface of Earth's processes, is fundamentally wrong. In his Nobel Prize acceptance speech, he said: "The experiences of the early 1970s had made it utterly clear to me that human activities had grown so much that they could compete and interfere with natural processes."

This far-reaching notion grew in Crutzen's scientific mind until it burst onto the scene in early 2000. In February of that year our then sixty-seven-year-old scientist went to Cuernavaca, Mexico, to take part in an International Geosphere-Biosphere (IGBP) conference, a forum for Earth system research, hosted by the United Nations. The debate revolved around human impacts on the environment and, time and again, the term for the geological epoch in which we live came up: the Holocene. The Holocene is said to have started 11,700 years ago, as the last Ice Age came to an end. Crutzen remembers the moment thus: "The chairman mentioned the Holocene again and again as our current geological epoch. After hearing that term many times, I lost my temper, interrupted the speaker and remarked that we are no longer in the Holocene. I said that we were already in the Anthropocene. My remark had a major impact on the audience. First there was silence, then people started to discuss this."

The Australian climate researcher Will Steffen describes the scene like this: "Scientists from IGBP's paleo-environment project were reporting on their latest research, often referring to the Holocene, the most recent geological epoch of Earth history, to set the context for their work. Paul, a Vice-Chair of IGBP, was becoming visibly agitated at this usage, and after the term Holocene was mentioned again, he interrupted them. "Stop using the word Holocene. We're not in the Holocene any more. We're in the . . . the . . . the . . . (searching for the right word) . . . the Anthropocene![10]

10. Quoted from a commentary by Will Steffen on Paul J. Crutzen and Eugene F. Stoermer, "The 'Anthropocene' (2000)," in Libby Robin, Sverker Sörlin and Paul Warde (eds.): *The Future of Nature*, (2013): Yale University Press as well as personal communication with Will Steffen.

The word landed among the experts like a time-bomb. *Anthropos*: the Greek word for "humans," *cene*: from *kainos*, the Greek word for "new," *Anthropocene*: the new epoch of humans.

In the coffee break after the session, this new word was virtually the only topic of conversation. Crutzen had just redefined the context in which humanity exists on Earth. With it, he had portrayed everything humans do to and with Earth, normally measured in days, years and centuries, in a whole new way. Crutzen suggested a geological scale, of thousands and even millions of years. He had asserted that human activity has affected the Earth, on a geological scale.

The scientists in that conference room in Mexico were profoundly shaken because the Nobel Prize Laureate for Chemistry—one of the most often cited natural scientists in the world—was not only describing the *past* with this new term (something to which geologists are accustomed) but he was also redefining and connecting to the future of a world that is only just emerging: a new Earth sculpted by humans.

From the perspective of the Anthropocene, the ozone layer story will be just one of a hundred or a thousand ways in which humans are fundamentally altering this planet, 4.6 billion years after its formation. Until Crutzen's statement in Mexico, we had seen all this mainly from a narrowly short-term human perspective, and for the most part, we were unconscious of the consequences our actions had for the globe.

ONE Welcome to the Club of Revolutionaries

WHETHER YOU TAKE A WALK in the hills around your town or along the coast or by a river, you will encounter the results of geological forces that have been at work for millions of years. Magma that once was deep inside the earth has formed rocks and moved tectonic plates. Water has shaped shorelines and carved out deep valleys. Wind erosion has flattened mountains and created massive deposits of soil and sand.

The exact spot on the earth's surface that now lies beneath the city of Berlin, the German capital, where I live, was once near the earth's southern pole some 500 million years ago. Tectonic forces moved it north over that immense period of time.[11] Only tens of thousands of years ago, the area was covered with huge glaciers; the weight and the power of their melting water created today's landscape. Without much effort, one can also observe more recent changes due to geophysical forces. I only have to walk 500 yards from home to reach Heidelberger Platz, from where a wide boulevard runs toward the stores and cafés in the center of West Berlin. The difference in slope between one end of the street and the other is so slight that, in this otherwise flat city, most cyclists and motorists barely notice it. But there is a big story behind this slight slope. It was once the bank of a gigantic river that flowed here at the end of the last Ice Age, 12,500 years ago. This *Urstrom* (glacial river) was filled with icy water several hundred

11. Documentation for Berlin's geographical spot having traveled from the South polar region to its current location may be found in several sources: Stampfli, Gérard M., Jürgen F. von Raumer & Gilles D. Borel, "Paleozoic evolution of pre-Variscan terranes: From Gondwana to the Variscan collision," *Geological Society of America Special Paper* 364, 2002 and in Cocks, L.R.M. and T.H. Torsvik, "European geography in a global context from the Vendian to the end of the Palaeozoic," Geological Society, London, Special publications, 2006.

meters high, to the north of what is now Berlin, before the great thaws set in and the glaciers gradually melted away.

When I cycle down this slope, I hear cars thundering past. I try to imagine the thundering mass of water that used to rush past, which formed the landscape of sand and stone on which the city of Berlin arose in the thirteenth century. The opposite bank of this primeval river is almost six miles away in Prenzlauer Berg, one of Berlin's hip new districts. The river must have been gigantic and would make today's River Spree, which runs through the political and cultural center of Berlin, near the Brandenburg Gate, seem like a mere creek.

When you contemplate Earth's history—not just by rattling off things you learned at school but by touching stones or letting sand run through your hands or swimming in a river—even a brief encounter can turn into a fantastic adventure. For me, the excitement is even greater when I become aware of the workings of earlier life forms. Many inland hills found on continents are in fact the remains of ancient coral reefs. Many mountain ranges far from the sea are composed of the calcareous skeletons of earlier marine organisms. Thick deposits of coal and oil, which have provided the fuel for industrial prosperity, are the residues of earlier life forms. Here in Berlin, there is a lot of bog and marshland. When you go hiking where fauna and flora are scant, you sometimes feel as if you are in a Zen garden where lots of decaying moss is underfoot; if it were left undisturbed, these mosses would eventually form coal. In bogs like these you can witness geology at work. You can see how the stones here and the earth's crust are connected to life itself.

Earth's surface, as we know it today, has been transformed by a select group of organisms which I refer to as "The Club of Revolutionaries." These are the life forms that did not die out unceremoniously after a mere couple of million years. These are the species that did not just surrender their molecules to the great recycling process called evolution, to be absorbed by other life forms.

The Club of Revolutionaries is comprised of species that have caused lasting change and have created new structures, just as fire, water and wind have done. We still encounter them, eons after their biological demise, in the form of bizarre limestone sculptures, or as pitch-black coal seams, deep below the ocean.

The oldest—and from our point of view, most essential "revolution-ary" is the one that has made possible today's earth, with all its trees and flowering plants, birds and mammals. This revolutionary is a tiny micro-organism that has evolved over three billion years. It used to be called blue-green algae but this label was discarded once scientists realized they weren't dealing with algae at all but rather with bacteria. Since then, such life forms have been referred to as cyanobacteria. They paved the way for life to use the sun's energy and to spread from sea to land across the whole surface of the planet.

Before cyanobacteria entered the scene, a young earth, amassed from matter orbiting the sun, had already been through some dramatic changes. It had been hit by another celestial body, a space traveler roughly the size of Mars.[12] *Theia*, as it's now called, created such impact that the moon was ejected from the earth's mass. As a result, the planet's axis of rotation became tilted, leading to tides and seasons. The fiery interior of the earth still holds the heat from that impact—so, in a sense, we don't live on one planet, but actually two. After earth's and Theia's matter had merged, a core formed, composed mostly of iron, and a new magnetic field developed, shielding the planet's surface from harmful radiation from space. Next, a primordial atmosphere began to coalesce, consisting of toxic gases that would certainly be fatal to contemporary organisms. And then, approximately 3.7 to 4 billion years ago, a second "Big Bang" occurred, this one biological. Simple molecules morphed into cells that could replicate themselves. The earth now began to sustain life. In con-tinuous cycles of mutation and replication, adaptation and extinction, these first life forms, now known as archaebacteria, evolved. But they were soon to be confronted with an early resource crisis. The chemical energy they needed for survival became increasingly scarce in their primeval world.

It was then that cyanobacteria entered the scene. Their altered metab-olism proved to be superior in one essential respect: whereas archaebac-teria were dependent on the earth's chemical energy, cyanobacteria were

12. Alex Halliday, "The Origin of the Moon," *Science*, vol. 338, no. 6110 (2012): 1040–1041; Matija Cuk and Sarah Stewart, "Making the Moon from a Fast-Spinning Earth: A Giant Impact Followed by Resonant Despinning," *Science*, vol. 338, no. 6110 (2012): 1047–1052.

able to tap into the sun's constant flow of energy. They developed molecular networks and metabolic pathways—the ability to convert energy from light and heat to enable small cell photosynthesis. Thus life's first resource crisis was solved to its advantage, yet if viewed from archaebacteria's perspective, it also created the first environmental disaster. Photosynthesis generated large quantities of oxygen. This element had already been present in the earth's atmosphere in its poisonous molecular form, O_2, but only in limited quantity as a trace element.

Now, cyanobacteria were pumping large amounts of O_2 into the atmosphere. Over the course of millions of years, the concentration of this gas grew, with far-reaching consequences. For archaebacteria, oxygen was poisonous, so they retreated to very remote locations, like deep-sea vents. Cyanobacteria, on the other hand, fared so well in this new oxygenated world that they multiplied, eventually spreading across the oceans and coastal regions, to form extensive mats and vast nodular colonies.

Thus, cyanobacteria became founders of "The Club of Revolutionaries," They released so much oxygen into the atmosphere that around 2.6 billion years ago, dissolved iron in the seas began to oxidize and settle to the bottom. Vast deposits of iron ore were formed, used today in the construction of buildings, complex machines and electronic equipment.

Once the oceans were saturated with oxygen, surplus gas escaped into the atmosphere, and the next revolution began. High up in the sky, ultraviolet radiation transformed some of this copious O_2 into O_3. (O_2, which contains two atoms of oxygen, is much more stable than O_3, with its three oxygen atoms.) This transformation created the ozone layer, which has intercepted the most aggressive radiations from the sun. (That is, until a life form called Thomas Midgley began tinkering with artificial chemical compounds). It was only due to this protective layer around the "sea of air"—as Alexander von Humboldt called the atmosphere—that new, more complex life forms could evolve. Approximately 420 million years ago life, in the form of plants, amphibians, reptiles and mammals, spread over the land.

Cyanobacteria not only provided these more complex life forms with the oxygen necessary to digest food effectively, they also passed on the molecular technology to produce it. According to a widely supported hypothesis, all multicellular plants came into existence by absorbing cyanobacteria and using them as interior "solar panels" to generate photosynthesis.[13] Cyanobacteria thus became a component of each of the quarter million plant species known today, from cacti and dandelions to Sequoia trees. They have even stored a ration of their own genetic material. For these partners, it was a win-win situation. Cyanobacteria's distant descendants are found everywhere plants grow, making the forest green. In addition, free-roaming cyanobacteria are still around. Two thousand contemporary species have been recorded to date. In the 1980s, American marine biologists Sallie W. Chisholm and Robert J. Olson, along with other collaborators, discovered a life form that had been previously overlooked. The organism was tiny but once it was detected, further research revealed that the cyanobacteria *Prochlorococcus marinus* was one of the most common organisms on earth and was probably one of the most widely spread types of picoplankton in the world.[14]

Most people today are unaware of cyanobacteria except in unpleasant circumstances. If they are present in large quantities, due to fertilizer runoff and warm weather, they can produce substances that irritate human skin. But in places like Australia, cyanobacteria can also be admired: For millions of years, they have formed large colonies where their excretions produce stone-like structures, called stromatolites.

No matter where you are and what you do, when you breathe to stay alive or enjoy time outside, when you eat vegetables or buy something made of iron or steel, you are inextricably linked to these revolutionaries.

This extraordinary feat surely merits having a memorial erected in every modern city, in honor of the founders of the Club of Revolutionaries: "To the creators of the oxygen atmosphere, our planet's protective shield, the plant world and iron deposits: In gratitude, humanity."

So far, that hasn't happened. But in the step-by-step process of science, humanity is at least starting to discover how deeply connected we are, not only to our primate ancestors but also to a whole set of life forms that

13. See seminal article of Lynn Sagan, "On the origin of mitoting cells," *Journal of Theoretical Biology*, vol.14 no.3, March 1967.

14. Sallie W. Chisholm et al., "A novel free-living prochlorophyte abundant in the oceanic euphotic zone", *Nature*, 1988, vol. 334 (1988): 340–343 and F. Partensky *et al.* "*Prochlorococcus*, a marine photosynthetic prokaryote of global significance", *Microbiology and Molecular Biology Reviews* vol. 63 (1999): 106–27.

have made and continue to make earth livable. By doing research, humans have learned how bacteria, plants and animals have sustained life on earth and they have even begun doing experiments that attempt to recreate the conditions by which earth has stayed habitable.

One of the first to do this kind of research was Joseph Priestley, a British chemist, theologian, philosopher and physicist. In 1772, he founded the discipline of earth modeling. Today, earth modelers have the advantage of gleaning reams of data from satellites and supercomputers. Priestley, who was interested in oxygen and who is regarded as one of its discoverers, worked with simpler technology. He trapped mice under a bell jar and watched what happened. After disposing of the inevitably dead animals several times, he was surprised when he observed that mice survived if he included a green, living plant, thus creating a tiny, enclosed ecosystem.

Priestly wrote the first ever description of photosynthesis, describing how animals and plants interact, in his inimitable prose: "These proofs of a partial restoration of air by plants in a state of vegetation, though in a confined and unnatural situation, cannot but render it highly probable, that the injury which is continually done to the atmosphere by the respiration of such a number of animals, and the putrefaction of such masses by both vegetable and animal matter, is, in part at least, repaired by the vegetable creation."[15]

With his bell jar, Priestley inspired a whole new research discipline: ecology, and later biospherics, the study of artificial, enclosed ecosystems. In 1875, Austrian geologist Eduard Suess created the term "biosphere" to describe the space used by living organisms. A few decades later, the Russian geologist Vladimir Vernadsky expanded this concept when he realized that the biosphere is not only inhabited by living organisms but has also been shaped by them. Vernadsky demonstrated how humans are existentially a part of the biosphere.

When both the USA and the USSR were in a race to reach the moon and conquer the vastness of space, Russian scientist Yevgeny Shepelev confined himself in the smallest possible artificial ecosystem, assigning himself the role of Joseph Priestley's mice.

15. Joseph Priestley, "Observations on different Kinds of Air," Philosophical Transactions of the Royal Society, 62, (1772): 147–264, quoted from Malcolm Dick (ed.), *Joseph Priestley and Birmingham*, Brewin Books (2005).

In 1962, he climbed into a small, airtight metal container at the Institute of Biomedical Problems in Moscow. When he sealed the door behind him, he was not alone: he shared the cramped space with forty-five liters of green algae, of the genus *Chlorella*. His plan was for the algae to supply him with the oxygen he needed to survive. This forty-two-year-old Russian was the first person to make himself completely dependent on a bucket of plants.[16]

Shepelev grew up with eight siblings in impoverished circumstances. He discovered his love of science very early in life and managed to be accepted into the scientific youth club at the Moscow Zoological Gardens. He then studied medicine and devoted himself to a broader subject: how life could survive in outer space. He wanted his containers to show that cities of the future could be built and maintained, on other planets. Thus, the Soviet Union would colonize outer space before the capitalist West.

Shepelev's first experiment lasted a mere twenty-four hours. When a colleague opened the door of the container, he complained of being hit by a rotting smell. Its occupant was dazed and confused, his thought processes befuddled by his own exhaled gases. Yet, Shepelev had actually managed to live off the oxygen produced by his algae.[17]

In the Siberian city of Krasnoyarsk, other scientists were undertaking similar, strictly confidential research. In 1972 three scientists managed to survive for half a year in Bios-3, an artificial ecosystem, without external supplies of water and oxygen. By the end of the 1980s, Russian scientists succeeded in producing three quarters of the food they needed, in "closed" systems. Since a diet consisting entirely of algae made them feel bad tempered, they started growing cucumbers, tomatoes, potatoes, peas and other container plants, and even created a new type of soil that was dubbed "soil-similar substrate."

In time, the Russians became more daring. When the political climate in the Soviet Union began to change in the mid-1980s, they even started doing tests to measure environmental problems that, by definition, did not exist in

16. The biographical details were obtained from the Institute from Biomedical Problems in Moscow in a personal communication, April 2010.

17. Personal communication with Prof. A.G. Degermendzhi, Director of the Institute of Biophysics and Prof. A.A Tikhomirov, Director of the International Closed Ecosystems Center in Krasnoyarsk, April 2010.

a "socialist" society, even though everyday Soviet life was full of them. The scientists pumped pollutants into containers to investigate the effects. "The ability to buffer these kinds of substances and transform them is limited," stated a terse summary in one report. What was meant by this was clear: ecosystems can handle stress for a long time but under continual stress they will eventually collapse.[18]

While the Russian experiments were being carried out in secrecy, a group of scientists and idealists in the USA, meanwhile, were working on a significantly more complicated artificial ecosystem. In 1991, with great fanfare, Biosphere 2 was inaugurated by John P. Allen, a maverick with a background in mining and metallurgy, who had a strong personal vision that Biosphere 1, planet Earth, was in big trouble. Allen once held the rights to a coal seam worth a fortune, but according to him, the expected course of his life completely changed after a transformative experience with the psychotropic plant peyote which made directly aware of the biosphere.[19] Consequently, in the late 1980s, he formed an unlikely alliance between ecologically-minded friends and associates, American scientists and the Texan venture capitalist Edward Bass, in order to build the largest self-contained ecosystem in the world.

Between 1991 and 1994, two groups of "biospherians" lived in an enormous, sealed glass, cathedral-like structure, in the Arizona desert, which had taken four years to build. The first crew of eight spent two years inside, Biosphere 2 was a manifestation of research, environmental education and media hype. Like Yevgeny Shepelev, both John Allen and Edward Bass were interested in future settlements in space. Biosphere 2 may have looked like a study of how humans can live in an artificial environment but it turned out to be the complete opposite.[20]

The Biosphere 2 project generated a great deal of interest worldwide. For an admission fee, visitors were even allowed into Biosphere 2, itself. The grandiose white structure housed a man-made rain forest, an ocean,

18. Frank B. Salisbury et al., "Bios-3: Siberian Experiments in Bioregenerative Life Support," *BioScience*, vol. 47 (1997): 575–585.

19. John Allen has written an autobiography: John Allen, *Me and the Biospheres*, Synergetic Press, Santa Fe, NM, 2009.

20. John Allen et al., "The Legacy of Biosphere 2 for the study of Biospherics and closed ecological system," *Advances in Space Research*, vol. 31, no. 7 (2003):1629–1639.

a coral reef, a mangrove swamp, a desert and a savannah, all in minia-ture forms. Two and a half thousand square meters of agricultural land were set aside to produce food for the biospherians and a diverse selection of animals, ranging from bees for pollination to pygmy goats, were also included.

The first crew of eight biospherians moved into the enormous complex in 1991. Their aim was to live in the synthetic ecosystem for at least two years or for as long as possible, researching and observing how conditions for life changed over time. Serious problems arose, early on. The concen-tration of oxygen in the air dropped continuously during the first year and a half, from nearly 21 percent atmosphere, as in Biosphere 1, to 14.5 percent, similar to mountain air at four thousand meters above sea level. It took a while to ascertain the cause: Bacteria in the virgin soil, was consuming oxy-gen in the air while at the same time chemically active concrete walls were absorbing oxygen and producing calcium carbonate. Fatigue set in among the biospherians, so measured amounts of liquid oxygen were pumped in. The food supply was erratic, too. Pollinating insects died off while ants and cockroaches thrived. The harvest produced a smaller yield than expected, partly because of two consecutive years of the El Niño weather phenome-non, so reserve food stocks had to be used. The first mission ended after the prescribed two years setting world records (by far) for duration in enclosed human life support experiments. A second crew entered Biosphere 2 in March 1994 but the mission was terminated prematurely as a result of dis-putes between Allen and his design/management team, with his partner, Ed Bass and his team. After the partnership dissolved, the goals of the project shifted away from human enclosure experiments.[21]

While many in the media suggested it was a failure, Biosphere 2 had made an incredible contribution, not least with the many research and sci-entific papers it has produced and indeed is still producing.[22] It is because of the difficulties the project encountered that it became even more significant.

21. Personal communication with John Allen.
22. Most of the key published papers available at www.globalecotechnics.com. Elsevier special edition: *Biosphere 2 Research Past and Present*, eds.. B.D.V. Marino, H.T. Odum, Ecological Engineering Special Issue, Vol. 13, Nos. 1-4, Elsevier Science, 1999.

Each problem with living in an artificial ecosystem symbolizes the present situation of humanity. The scores of deriders who made fun of the bionauts in Russia and the biospherians in the Arizona desert must have forgotten how much harm people in the real world cause to the ozone layer, or to precious animal species that could become extinct before our very eyes. People forget what causes a shortage of food supplies for nearly a billion people or how we risk making the earth's climate very uncomfortable for ourselves.

The Russian and American projects yielded an essential insight: the earth is constantly providing us with a multitude of services and processes that have evolved over the course of hundreds of millions of years, thanks to the work of early earth "revolutionaries." If you want to re-create these services in the form of huge artificial ecosystems that can sustain hundreds of millions of people, the costs will clearly rocket into infinity. Even 150 million dollars was not sufficient for the Arizona experiment to sustain eight people in a 1.2–hectare artificial ecosystem. These biospheric projects therefore showed that nature sustains human civilization and the world economy.

Today, the University of Arizona owns and directs research at the Biosphere 2 facility. It would be good if there or elsewhere, bionauts or biospherians would again move inside enclosed systems to determine if humans can survive in strictly confined spaces.

It is telling how much people appreciate the oxygen created by cyanobacteria or simple plants or fruits when they are cut off from nature. In November 2013, Japanese astronaut Koichi Wakata tweeted an image from the International Space Station. It showed a tomato in a state of weightlessness, while the earth could be seen in the background. "One fresh tomato for dinner makes us happy in space. It came up with us on Soyuz TMA-11M, two weeks ago," read his text about the red marvel, seemingly appearing in front of the Blue Marble.[23]

In the Anthropocene, the earth itself becomes one giant biospheric experiment, but without any emergency exits or windows to let in additional fresh air. So, when you take your next walk outside, look closely, not only at the results at what wind, fire and water have carved out and

23. http://wordlesstech.com/2013/11/26/tomato-space/

what other organisms have left behind, but also examine the results of thousands of years of human activity. These cumulative actions stack up to look like a new geological epoch that puts us on a par with the cyano-bacteria and other earth-transforming species: Welcome to "The Club of Revolutionaries."

TWO **The Long March**

I n January 2013, American biologist and author Paul Salopek set off
to trek around the world. He began his journey at a lake in the Great
Rift Valley of Ethiopia, where remains suggest that modern man originated. Salopek's goal, over a period of seven years, is to walk all the way to
Tierra del Fuego— a region that lies the furthest from humanity's birth-
place—via the Middle East, Southeast Asia, China and North America.
His planned hike is about 30,000 kilometers long (21,000 miles); he hopes
to complete it in 2020. Salopek wants to retrace the path that humans have
taken since starting out from Africa.

Before humans came on the scene, many animal species had already
spread across the earth's surface. Humans, however, were the start of something quite new. The biggest difference between us and other members of
the Club of Revolutionaries such as cyanobacteria or algae is that we are
able to act consciously with the help of a molecular electronic mirror image
of ourselves and the world.[24] This difference, our uniquely precious ability,
is what makes the Anthropocene completely new: human consciousness
and geology forming a unity. By rights, there should be a UNESCO world
cultural heritage site in East Africa named "The Origins of the Human
Consciousness," where Salopek's journey started. The dimensions of this
evolutionary event are simply colossal, extending from the first groups of
humans who used pronouns like *I, you* and *we*, to online social networks
connecting billions of people. They link the earliest humans who looked in

---

24. Recommended literature on the evolutionary history of consciousness: Giulio
Tononi, *Phi, A Voyage from the Brain to the Soul*, New York: Pantheon, 2012.

awe at the sky to the builders of the Hubble Space Telescope, and the first stone tools to quantum computers of the near future.

Part of our emerging consciousness is an ever increasing awareness of our long march, how deeply rooted we are in the cosmos, the solar system, our planet and life, This awareness is not always present in our everyday lives, what with washing dirty dishes, chatting on social networks, or meeting deadlines. Part of the appeal and beauty of scientific research is to develop this awareness more thoroughly in ever greater detail.

Humans are the way we are because water, carbon, oxygen, minerals and metals were distributed across earth from the beginning in very specific proportions. We are the result of hundreds of millions of years of evolution, from individual unicellular organisms that began to cooperate, to the estimated 37 trillion cells combining to assemble organs as complex as the human brain.[25]

Not only every stone but every person encapsulates the entire history of life and the universe.[26] The atoms making up our bodies have been traveling through the cosmos for billions of years,. Each atom in our bodies has already served to build hundreds of other life forms before us; perhaps it swam in a fish long before humans appeared or lay deep in the soil or was a building block in a bacterium. Our bodies are gigantic zoos of evolutionary history. Hundreds of millions of years of evolution has shaped the way we think and perceive, from the first four-celled organisms with nerve endings on to fish, and from there to the first four-legged animals to primeval mouse-like mammals, on to the first primates who developed into early humans. After millions of years, a species of thinking, humanoid life forms rose to the challenge of survival in the vastness of Africa and developed characteristics similar to what we humans have today.

Modern-day people have brains that have been transformed by environment, long before people acquired the ability to change the environment. Nowadays, what we like or dislike, or fear or do not fear, or perceive

---

25. Eva Bianconi et al., "An estimation of the number of cells in the human body," *Annals of Human Biology*, 5 July 2013.
26. Excellent further reading on this subject in: Jan Zalasiewicz, *The Planet in a Pebble, A Journey into Earth's Deep History*, Oxford University Press, 2010.

to be or not, have much to do with the living conditions of our ancestors. These forebears include the squirrel-like *Purgatorius* that lived soon after the extinction of the dinosaurs; *Eosimias*, one of the earliest anthropoids that lived 41 million years ago; the chimpanzee-sized *Kamoyapithecus* that lived 25 million years ago, regarded by many researchers as the first hominoid and the last mutual ancestor before chimpanzees, that lived in Africa approximately 7 million years ago. This leads finally to the first hominids such as *Sahelanthropus* and *Australopithecus* that paved the evolutionary path for the genus *Homo*. Our lives today are linked by invisible threads to this past; each set of respective environmental circumstances, from the meteorite impact that doomed the dinosaurs, allowing for the era of mammals (who could otherwise have become the pets of highly intelligent dinosaur descendants) to the expansion of the savannah in East Africa due to natural climate change.

Besides our consciousness, what makes us humans uniquely able to create the Anthropocene, is our incredible generalism, that is, our ability to adapt not only to new circumstances but also to be the shapers of our habitat.[27]

Our biological constitution is, for the most part, an echo from the past three million years when the earth was significantly colder than it is today. A factor that contributed to this cooling process was the formation of the Isthmus of Panama three million years ago, which connected North and South America and interrupted the flow of warm water from the Pacific Ocean over to Africa. Atlantic currents were forced northwards, eventually leading to the formation of today's Gulf Stream. The Himalayan mountain range also continued to rise, which rerouted Asian rivers to flow northward rather than south. Flowing into northern seas diluted their salt concentration: water that is low in salt freezes more quickly, which led to the glaciation of the Arctic region. The sea level sank during these periods to an average of 426 feet because the water froze.

Then the level rose again as subsequent oceanic and atmospheric changes led to warming periods. The Great Rift Valley in East Africa, where early hominids lived, rose slowly but continuously during this epoch,

27. See also Erle Ellis' article "Conserving a used Planet: Embracing our History as transformers of Earth," *Snap Magazine*, http://www.snap.is/magazine/embracing-our-history-as-transformers-of-earth/.

creating a drier habitat and causing forests to change to savannahs—an environment where an upright gait was a significant advantage.[28]

Early humans lived during times of sweeping change in the natural world that surrounded them. Whereas many species remained unchanged and only reacted to events, our ancestors were more flexible, innovative, and adaptable. Environmental changes even tended to foster flexibility and generalism in early humans. They developed an ability to survive under varied conditions, sustained by a wide variety of foods. The route to today's world, wherein people can live in Arctic cold or tropical heat, on mountaintops or in river deltas, in Indian slums or in air-conditioned New York stockbrokers' offices, arose a good two million years ago; omnivorous hominids proving to be very skilled at adapting. Due to genetic changes, brain size increased more rapidly, something that did not occur in close hominoid relatives whose numbers were much greater. These creatures began to use stone tools and began the evolutionary journey towards Homo; our branch on the tree of life.[29]

That branch spread north two million years ago, from East Africa toward the Mediterranean and from there into Asia, even as far as present day Indonesia and China. These prehistoric peoples wandered only a few miles each generation, eventually reaching Europe where, as far as can be determined, they lit the first fires, during the cold era that occurred about four hundred thousand years ago. Neanderthals were one of the first waves of this human expansion.

But a more significant revolution, a human "Big Bang," also took place in Africa, about 220,000 years ago—a mere instant in geological terms—a lighter, more agile creature called *Homo sapiens* emerged. Everyone alive today is related to this "new kid on the block." He is the most social yet egotistical, loving yet cruel, sensible yet emotional, far-sighted and narrow-minded, creative and destructive of all hominids. Our ancestors survived dangers and setbacks, and began a triumphal march out of Africa across the globe, the march Paul Salopek is following with his "Out

28. For a general depiction of the history of the climate, see Jan Zalasiewicz and Mark Williams, *The Goldilocks Planet—the four billion year story of the earth's climate*, Oxford University Press, 2012.

29. For a comprehensible description of human evolution, see Alice Roberts, *Evolution —The Human Story*, Dorling Kindersley, 2011.

of Eden" project. A hundred thousand years ago, they settled in what is now the Middle East, seventy thousand years ago they arrived in Australia, about forty-four thousand years ago they came to Europe and entered the habitat of the Neanderthals and about thirty thousand years ago they came from the north, moving into the entire American landmass.[30]

The spread of *Homo sapiens* had disastrous consequences for Neanderthals, human beings with artistic, cultural and even religious sensitivities. In their competition for land and resources, Neanderthal humans drew the short straw. Early *Homo sapiens* had already wiped out countless other species of animals—mostly large predators and some species they most enjoyed eating—a mere foretaste of the Anthropocene wave of extinctions to come, perhaps. According to recent findings, our direct ancestors cannot be held responsible for the extinction of the Neanderthals since there are no signs of massacres or widespread slaughter.[31] There are even strong indications that Neanderthals and *Homo sapiens* had sex and that some Neanderthal genes persist today. But that didn't stop Neanderthals from going extinct. It may have been enough for our ancestors to be just a little more efficient at hunting and gathering in shared regions and in using up the resources that Neanderthals needed to survive. Thirty-seven thousand years ago, the trail of this fascinating alternate species of human disappears, whereas the spread of *Homo sapiens* truly kicked off.

The next decisive point in humanity's ascent happened about twelve thousand years ago. The end of the last Ice Age and the beginning of a natural global warming created ideal conditions for a truly global expansion. Human ingenuity, fertile soil and a more favorable climate coalesced in a unique way. Independent of one another, human groups abandoned nomadic life and became agriculturalists, settling in fecund regions of the world, like the "Fertile Crescent," the Andean Altiplano, Mesoamerica, China and New Guinea.

30. On the first modern human in Europe, see Stefano Benazzi et al., "Early dispersal of modern humans in Europe and implications for Neanderthal behaviour," *Nature,* vol. 479, no. 7374, November 2 (2011): 525–528. On the first modern humans in Australia, see Morten Rasmussen et al., "An Aboriginal Australian Genome Reveals Separate Human Dispersals into Asia," *Science,* vol. 334, no. 6052, October 7 (2011): 94–98.

31. Dálen, Love, "Partial genetic turnoverbin neandertals," *Molecular Biology and Evolution,* February 23, 2012.

Some of these early farmers settled in an area comprising modern day Azerbaijan, Iran, Iraq, Turkey and Syria. They discovered that grass seeds are not only edible but can also replicate. Precious calories could be gained from this food source, supplementing the hunting of gazelles and the gathering of nuts and berries. Once under way, the agricultural revolution could not be stopped.

Farming had some grave disadvantages: each calorie yielded from the land took more time and effort compared to hunting for meat.[32] Farmers had less free time than hunters and gatherers. Yet, farmers found it worthwhile to work the land since it enabled them to stock up food in case of hard times. Surplus food meant that more children survived their first few months of life but also there were now more mouths to feed.

Layer upon layer, this history is evident at the village of Abu Hureya, an ancient settlement on the upper reaches of the Euphrates River, in present-day Syria. British archaeologist Andrew Moore examined the village before it disappeared under floodwaters caused by the construction of Syrian president Hafez al Assad's massive new dam.[33] People of the Natufian culture lived here around twelve thousand years ago. Remains of hundreds of different plants have been discovered beside the residents' simple homes. Einkorn and emmer wheats, rye, lentils and fava beans play a central role in life. These people were very innovative: they discovered ways of maintaining the fertility of the land, by planting pulses which return nitrogen to the soil, and they domesticated animals such as goats, sheep and later cattle, brought down from the nearby Zagros Mountains. The symbiosis between humans and farm animals had begun to develop.

We have now arrived at a critical moment in our high-speed review of human history. The Holocene is the period of earth's history in which we currently, officially live, based on geological calculations. Before modern humans were the children of the Holocene, our closest ancestors inhabited the Pliocene, a geological epoch that began 5.3 million years ago and ended 2.6 million years ago when the Pleistocene era began. If the Pleistocene Ice Age had simply continued, it is conceivable that humans would have

32. David R. Montgomery, *Dirt—The Erosions of Civilizations,* Berkeley: University of California Press, 2007.

33. Andrew Moore et al., *Village on the Euphrates: From Foraging to Farming at Abu Hureya,* Oxford University Press, 2000.

remained hunters and gatherers. But something "new" happened, which is what "Holocene" means (from the Greek, *holos*, whole or entire and *kainos*, new). As far back as seventy thousand years ago, our ancient relatives had already produced paintings on the walls of South African caves, and thirty thousand years ago, they had fashioned pipes from bones, made sculptures, needles, and ceramics. In many places, such as the Chauvet cave in the south of France, they created paintings that would rival those by Picasso or Franz Marc. Human are artists, masters at imagining, at creating, at reshaping their environment. Embedded in the favorable Holocene climate, these abilities have changed the world.

The start of the warming after the last Ice Age, approximately eleven thousand seven hundred years ago, prepared the conditions for "modern" life. Since the Holocene began, our biological make-up has changed very little. What *has* changed radically is our social, economic and technological make-up.

A few hundred agricultural pioneers in the Middle East have become a billion farmers who produce an inconceivable assortment of edible crops. The first fields and pastures have changed into a gigantic agricultural area of approximately 20 million square miles, which is larger than the entire surface area of the whole American continent. From scattered herds of sheep, goats and cattle, a global herd of livestock has grown, consisting of more than 50 billion animals, making up 90 per cent of the biomass of all the mammals on the Earth.[34]

Where once there were small villages, megacities have now grown; from the simplest tools, there are now coal excavators, 3D printers and plasma screens; from characters and symbols scratched on tablets, the World Wide Web. However, spears have evolved into missiles and combat drones. In amongst our anthropogenic burgeonings some very dark flowers have also sprouted.

If we fast-forward past the first cities, the culture of Imperial China, the great empires of antiquity, the development of global trade routes, the

---

34. See Gowri Koneswaran and Danielle Nierenberg, "Global Farm Animal Production and Global Warming: Impact and Mitigating Climate Change," *Environmental Health Perspectives*, vol. 116, no. 5 (January 2008):578–582 and Food and Agriculture Organization of the United Nations (FAO), "World agriculture–towards 2015/2030," Rome, 2002 and FAO and OECD, "Agricultural Outlook 2009–2018", Rome, 2009.

European conquest of the world, scientific breakthroughs and medical and technological progress—then the Holocene appears to be one extended, magnificent gift to humanity.[35]

No matter how tough the Holocene may have been for many people, it was characterized by boundless natural resources that could be discovered, extracted and utilized. Despite thunderstorms and weather extremes, earth's climate during the Holocene has been astonishingly stable, permitting us to build villages, towns and cities, and to farm. The last glaciation left behind wonderfully fertile soils like loess. Nature's services, by the thousand, providing water, soil or the air we breathe, have been available free of charge, without requiring any favor in return. Imagine if we were merely the second intelligent primate species and had to earn our living and obtain our resources in fierce competition with an entire civilization of *other* technology freaks, But we were lucky: the gold seams in California, the emeralds in India and the diamonds in Russia were all found untouched, for the first time, by humans. Our civilization has been one big treasure hunt.

All this came to a head in the eighteenth century, at another civilizational watershed, when people learned how to use the energy made available by earlier members of the Club of Revolutionaries: energy from the sun that led to the formation of coal and crude oil or natural gas. *This* is the moment Paul Crutzen suggests represents the transition from the Holocene to the Anthropocene—the emergence of humans as a veritable geological force.

When our great-great-great-grandparents discovered how to use this energy to power machines, humanity's potential increased, at a stroke. It was as if people had been given a collective potion that harnessed the strength of millions of horses and workers in the form of black chunks of coal and viscous oil. So much is taken for granted these days that we hardly notice. But if you've ever sweated to shovel a cubic meter of soil and then watched a backhoe do the same job, you too have experienced the Industrial Revolution, in one instant. Using fossil-based fuels and machines that could be powered by them, humanity *really* started to accelerate.

35. A very good overview of the ascent of human civilization in Asia and Europe can be found in Ian Morris's book with the slightly misleading title, *Why the West Rules for now—The patterns of history and what they reveal about the future*, Profile Books, 2011.

And this is why Paul Salopek's "Out of Eden Walk" confronted him with an entirely new reality just a few days after leaving the origins of humanity behind him: "Moving north and then east, we abandon the desert and stub our toes on the Anthropocene—the age of modern humans. Asphalt appears: the Djibouti-Ethiopia road, throbbing with trucks. We drift through a series of gritty towns. Dust and diesel. Bars. Shops with raw plank counters. Garlands of tin cups clink in the wind outside their doors. Then, near Dubti: a sea (no, a wall) of sugarcane. Miles of industrial irrigation. Canals. Diversion dams. Bulldozed fields. Levees crawling with dump trucks."[36]

36. Paul Salopek keeps a fascinating online journal about his project, see www.outofedenwalk.com.

THREE The End of the Holocene

A T THE BEGINNING OF THE INDUSTRIAL AGE, the side of the earth not facing the sun would have appeared completely dark when seen from space. Light from campfires, candles and oil lamps did not penetrate beyond earth's atmosphere. But then, people began to systematically draw upon the stored energy of the sun found in underground deposits to light their lamps and to power machines. From that time forward, fossil fuels have enabled the fascinating acquisition of material wealth. Since industrialization, one dot of light after another has shone from earth's dark side, like a long Promethean chain of lights, gas flames and burning forests: "It is like a running blaze on a plain, like a flash of lightning in the clouds. We live in the flicker—may it last as long as the old earth keeps rolling!" says the protagonist in Joseph Conrad's *Heart of Darkness,* referring to the lights along London's river Thames. (Significantly, Conrad locates the "heart of darkness" in London, not in the forests of Central Africa's Congo).

Over decades, these individual lights from houses, streets, offices, factories and burning forests have combined to form broad areas of illumination in intricate patterns that stretch along coastlines. We are sending a collective, moving sequence of lights into space, images of which get beamed back down to earth by satellites or by astronauts on the International Space Station. These images, some of which have been overlaid with cool musical tracks and posted on websites, inspire filmmakers like Alfonso Cuarón, the director of the movie *Gravity.*[37] In China, an urban

37. The short film of a circumnavigation of Earth by the ISS astronauts in the international space station is a must-see: http://vimeo.com/32001208.

mega-region over a thousand miles long is taking shape, while in West Africa, a coastal conglomeration nearly 600 kilometers (400 miles) in length is growing.[38]

A succession of technical, social and economic innovations has enabled people to completely change the face of the earth in a mere two hundred years, spreading themselves and their accomplishments across almost the entire planet.

Tapping into fossil fuels has had positive effects on the lives of billions of people today, in the form of hospitals and schools, global mobility and a mind-blowing array of consumer goods. The proportion of people living in abject poverty nowadays has also sharply declined. Educational opportunities, particularly for women, have greatly increased. Since 1990, life expectancy at birth has increased globally by six years, to an average of 70 years of age.[39]

The benefits of modern life are myriad: the use of a simple plastic cannula can save the lives of both mother and child at birth; driverless cars can take us, as if by magic, anywhere we want to go; research laboratories make it possible for billions of people to give free rein to curiosity. The late twentieth and early twenty-first centuries represent a time of incredible expansion of the human comfort zone, certainly for those who have the means and have never suffered the acute violence of war or the slow violence of poverty.

Without this new economy, without industrial-based agriculture or pharmaceuticals or fossil fuels, most of us would not be alive. Some people sound as if they would welcome a scenario of fewer people. The "population bomb" has been widely used as a metaphor. But in my opinion, the mere number of people is not the problem in the dawning Anthropocene. Being German, with the backdrop of my country's National Socialist past, I wouldn't want to imagine how a certain population number is considered "too many." Who is the perpetrator and who the victim in such a scenario? I think that every new person enriches earth with his or her

---

38. Karen C. Seto et al., "Global forecasts of urban expansion to 2030 and direct impacts on biodiversity and carbon pools", *PNAS*, August 16, 2012 http://www.ncbi.nlm.nih.gov /pmc/articles/PMC3479537/.

39. http://www.who.int/gho/mortality_burden_disease/life_tables/situation_trends_text /en/index.html.

potential for consciousness, creativity and community. I am not someone who would prefer only one billion people (instead of seven or eight) to be living on earth, nor am I like Stephen Emmott, who finds a world population of ten billion to be a terrifying vision.[40]

I believe that the Anthropocene idea can help people see themselves as active, integrated participants in an emerging new nature that will make earth more humanist rather than just *humanized*. It would be absurd if an idea named the "Anthropocene" were characterized by a negative view of humans!

But even the most positive attitude toward humanity cannot save us from having to face up to the enormous—literally earth-shattering—developments at the end of the Holocene. Our population numbers signal ever growing consumer demand, ever more areas of land claimed by people, ever increasing energy consumption with its consequences for the climate, and new influences on evolution. Attentive readers will already be familiar with some of these factors. But only when looked at as a whole, do they create the broad overview necessary to see how the Holocene is coming to an end and something new, the Anthropocene, is beginning. Our individual actions, multiplied by the number of people who are alive and make decisions, is a new reality that is hurtling towards us with such velocity that its consequences, both positive and negative, surprise us.

If your head starts spinning at the huge numbers being mentioned here, just remember that these figures derive from the totality of many small actions. Millions of tons of eroded soil start with the food harvested from one industrially farmed field. Billions of tons of carbon dioxide emissions start with the flick of a switch, whether to turn on a light or a car engine. All the phenomena of the Anthropocene—whether positive or insane, surprising, funny or creative—start with small actions. When you buy a ballpoint pen that has a tiny, man-made crystal on the tip, you are thereby increasing the variety of Earth's minerals, something future geologists may wish to investigate. When you add another ton of carbon dioxide to the atmosphere, your descendants might swear at you long after you are gone.

There are four major factors determining the end of the Holocene. The first is population growth. If the number of people living today was

40. Stephen Emmott, *Ten Billion*, London: Vintage, 2013.

the same now as at the time of Jesus Christ—a few hundred million—their collective impact would not be sufficient to initiate a new geological epoch.

In the year 1800, there were one billion people; in 1930, two billion and in 1960, three billion. In October 2011, the seventh billionth person was born: Danica May Camacho of the Philippines was chosen by the United Nations to have this starring role.[41] If the world population was evenly distributed across the Earth, there would be fifty-three people per square kilometer of land (excluding the Antarctic).

By the middle of this century, according to United Nations forecasts, another two billion people will be added to the world population, which is equivalent to the number of people who were living on earth between World Wars I and II. This also means that by 2050, there will be about 140,000 more births than deaths, per day. By these calculations, a city with the population size of Los Angeles will be added to the world every month.

By the middle of this century, fifty-three people per square kilometer jumps to sixty-six people per square kilometer. All this growth is taking place mainly in developing countries. In other words, humanity is increasing every day by one Indian slum, one high-rise community in Beijing, one outlying district of Jakarta or one medium-sized town in the Congo.[42] Just the *number* of people on earth does not signify much: The greater impact comes from our way of life by which this number must be multiplied. Our consumption habits determine how *much* land, how *many* industrial and mining areas and how *much* urban space is necessary to sustain this number of people, not only their survival but also their happiness.

The second factor that marks the end of the Holocene is the enormous increase in human living space requirements. Only a quarter of the Earth (about 7,000 square kilometers or 12 million square miles) is arable land that is suitable for growing food for human consumption. By 2007, cities and communities had already extended over an area half the size of the

41. See http://www.theguardian.com/world/2011/oct/31/seven-billionth-baby-born-philippines.

42. Population Reference Bureau, *2009 World Population Sheet*, Washington, DC, 2009.

Australian continent, and this area is expected to grow considerably in the next decades.[43] Cities are really efficient at housing people, but nevertheless need a lot of energy and resources to be built and maintained. Gigantic quantities of concrete and other building materials are being produced, transported and deposited in order to create new settlements. Jan Zalasiewicz, a geologist at the University of Leicester, describes urbanization as "an alteration in sedimentation processes via the construction of man-made rock strata." Concrete is a key material in this process: "The global annual production is now approaching five billion cubic meters, that is something over two-thirds of a cubic meter for every man, woman and child on Earth, in total enough to cover all of Germany, Austria and parts of neighboring countries under a centimeter-thick layer of this stuff—each year. It is part of the urban stratum, rising above the ground surface as our homes and factories, and extending below it as foundations, metro systems, sewers, electrical cables, and yet deeper as mines and boreholes."[44]

Dams, mines and human induced erosion also comprehensively alter the geological state of the earth. Tens of thousands of hydroelectric dams stop enormous quantities of sediment from reaching estuaries.[45, 46] Erosion caused by industrial agriculture moves ten times the volume of sediment than was the average 500 million years ago.[47] Material flows of such essential elements as phosphorous and nitrogen, both of which are used in artificial fertilizers, are caused by human activity. With the production of artificial fertilizers by means of the Haber-Bosch process, humans

43. UNFPA, "State of World Population 2007—Unleashing the potential of urban growth," 2007: 45, https://www.unfpa.org/webdav/site/global/shared/documents/publications/2007/695_filename_sowp2007_eng.pdf.

44. Jan Zalasiewicz, in Nina Möllers and Christian Schwägerl, "Anthropozän Natur und Technik im Menschenzeitalter," catalog for the eponymous exhibition at the Deutsches Museum, Munich, 2014.

45. James Syvitski and A. Kettner, "Sediment flux and the Anthropocene", *Philosophical Transactions of the Royal Society*, vol. 369, no. 1938 (2011): 957–975, and James Syvitski et al. "Sinking deltas due to human activities," *Nature Geoscience*, vol. 2, no. 10 (2009): 681–686.

46. As a visual aid, see also James Syvitski, "Humanity's Planet: Dams in the US 1800–2003, on Youtube.com.

47. Bruce H. Wilkinson, "Humans as geologic agents: A deep-time perspective," *Geology*, vol. 33, no. 3 (November 2004): 161–164 http://geology.geoscienceworld.org/content/33/3/161.abstract.

have already extracted more nitrogen out of the atmosphere than has ever circulated through the land ecosystems.[48]

But our space requirements extend well beyond the land surface. The Holocene oceans seemed inexhaustible, a boundless global ecosystem, 1.3 billion cubic kilometers in size, many times larger than all land habitats put together.[49] In a very short time, humanity's impact on the oceans has also become far-reaching, stretching thousands of feet down where atomic waste was dumped, new mining projects (such as the one off the coast of Papua, New Guinea) are pursued and new oil wells are drilled. Above all, our fishing fleets are transforming the oceans. There are one million large fishing ships worldwide and three million smaller boats.[50] These vessels track down schools of fish with the same sonar technology that American and Soviet nuclear submarines used during the Cold War. The hauling capacity of these ships has increased sixfold since 1970. Many of them use rollers that flatten the ocean floor, crushing every structure in which sea life can hide. Corals are the victim of trawlers with heavy harnesses. Deep-sea fishing damages the unique natural wonder of underwater mountains. This behavior is equivalent to hunters clearing entire forests just to catch a few deer. Yet, since 1970, yield per ship has fallen by two thirds. More and more ships compete for fewer and fewer fish. In the words of one ocean expert: "It's a race to our own destruction."[51]

When the UN Food and Agriculture Association (FAO) began to register catch quantities in the 1950s, they recorded 20 million tons. This was followed by a rapid increase to over 90 million tons of wild seafood caught. Since then, haul sizes have reached a plateau and have even contracted. This is not due to political restrictions but to the depletion of many stocks. In spite of larger, more powerful trawlers, there is nothing left to fish. The FAO classifies eighty per cent of fish stocks as being fully or excessively depleted.[52]

48. P.M. Vitousek, "Beyond global warming: ecology and global change," *Ecology*, vol. 75: 1861–1876.

49. James Syvitski and A. Kettner, "Sediment flux and the Anthropocene," op. cit.

50. Food and Agriculture Organization of the United Nations (FAO), "The State of World Fisheries and Aquaculture," Rome, 2012, http://www.fao.org/docrep/016/i2727e/i2727e00.htm.

51. FAO and World Bank, *The Sunken Billions. The Economic Justification for Fisheries Reform*, Rome/Washington DC, 2008.

52. Food and Agriculture Organization of the United Nations (FAO), "The State of

Even a natural disaster like a tsunami has a human dimension. The monster wave that hit Japan in 2011 was the manifestation of a powerful undersea earthquake. It devastated parts of the east coast of Japan and created a massive wave of man-made debris—houses, garbage, ships and containers—to be swept first inland, then out into the Pacific Ocean. The debris was tracked by the National Oceanic and Atmospheric Administration (NOAA), which regarded it as the largest occurrence to date demonstrating how debris disperses away from a single point.[53] The resultant islands of debris were not the only anthropogenic phenomenon. The tsunami wave triggered a disaster at the Fukushima nuclear plant, widely dispersing radioactive materials, including isotopes that will continue to emit radiation of various types for thousands of years to come.

Wild nature no longer exists on land or out at sea. According to analyses by US researchers, in cooperation with Google, during the period between 2000 and 2012, 2.3 million square kilometers (about 880,000 square miles) of "natural" forest, disappeared. Only around 800 thousand square kilometers (about 308 thousand square miles) have been replanted while the remaining areas have been turned into agricultural areas, residential areas or into wasteland.[54] The FAO alerts us to an alarming development in which cleared woodlands and even replanted forests are often monotonous, with no biological diversity. These monocultures cannot sustain indigenous peoples and render few ecological services.[55]

What remains of the wild is the result of human decision-making, such as when an area is perceived as being of lasting value and is then protected by the local population or by environmental organizations, or by a corporation that concludes that exploitation would not be profitable. Even

World Fisheries and Aquaculture," Rome, 2012, http://www.fao.org/docrep/016/i2727e/i2727e00.htm.

53. Tony Barboza, "No 'island' of tsunami debris floating toward US, NOAA says," *Los Angeles Times,* November 6, 2013, http://www.latimes.com/science/sciencenow/la-sci-sn-tsunami-debris-noaa-20131106,0,3159522.story.

54. Matthew Hansen et al., "High-Resolution Global Maps of 21st Century Forest Cover Change," *Science,* 15 November 2013, vol. 342, no. 6160: 850–853 and Betsy Mason, "Incredible High-Resolution Interactive Map of the World's Shrinking Forests," Wired Online, November 14, 2013, http://www.wired.com/wiredscience/2013/11/google-earth-deforestation/.

55. Jianchu Xu, "China's new forests aren't as green as they seem," *Nature,* vol. 477, 371, (September 21, 2011).

in places where people think they are in the wild, they often come across traces of civilization when they take a closer look. This frequently happens in Amazonia where, during the clearing of allegedly pristine rainforest, traces of earlier settlements are found.[56]

The Anthropocene marks the end of the illusion that "somewhere out there," there are gigantic, unexplored, untapped, unused regions, areas of untouched nature surrounding what is man-made. Geographers Erle Ellis and Navan Ramankutty from the University of Maryland have got to the heart of this. Using data from satellite photographs, they have determined that only 22 per cent of the earth's surface is still wilderness and only 11 per cent of photosynthesis activity takes place in these wild areas. The remaining area consists of agricultural, residential and industrial zones and other "anthromes," that is, areas marked by humans. These have replaced former biomes. "This new model of the biosphere moves us away from an outdated view of the world as 'natural ecosystems with humans disturbing them' and towards a vision of 'human systems with natural ecosystems embedded within them,'" states Ellis.[57]

Let yourself drift across the digital globe offered by Google Earth and similar services. Don't zoom in on your own apartment but go instead to areas in the world you do not know. Enjoy the unusual colors, shapes and mysterious structures. This used to be a perspective reserved only for gods. Then, such sights began appearing in expensively produced James Bond movies! Now, you only have to whip a small personal device out of your pocket to zoom down and see for yourself what it means to live on a planet shaped by humans. That green, dense forest—can you see the paths?

That wide, deserted plateau—can you see the open cast mine?

That sparkling blue coral reef—can you see the American military base?

That gray-brown gravel plain . . . Oops, it's a city!

56. Michael Heckenberger et al., "Amazonia 1492: Pristine Forest or Cultural Parkland?" *Science*, vol. 301, no. 5640 (September 19 2003):1710–1714, DOI: 10.1126/science .1086112.

57. Erle Ellis and Navin Ramankutty, "Putting people in the map: anthropogenic biomes of the world," *Frontiers in Ecology and the Environment*, vol. 6, no. 8, (2008): 439-447 and Erle Ellis and Navin Ramankutty, "Anthropogenic biomes," *Encyclopedia of Earth*, Cutler J. Cleveland (ed.), Washington, DC, 2009.

Those white dots in the sea off the coast—are they fishing boats?

From high above, these landscapes can look like complicated scriptures, cancerous tumors, works of art, geometric patterns, military parades, bacterial cultures, or even large gardens. It is a sensational sight in which millions of human decisions have been put together and displayed. Irish artist David Thomas Smith has created highly symmetrical photomontages of landscapes touched by humans. The title of his body of work is "Anthropocene."[58]

In addition to our increase in numbers and our growing imprint on the surface of the earth, the third characteristic that marks the end of the Holocene is our enormous energy consumption and its consequences for the global climate. World population has increased by a factor of 5.4 since 1860 but energy consumption has increased by a factor of 41, in the same period. On average, each individual now consumes the equivalent of half a gallon of petroleum—per day.[59]

The signature of humans is becoming visible on land in the form of shale fracking, in the oceans in the number of deep-sea drilling rigs and in the sky with a wide range of new chemicals in the atmosphere. From regional phenomena, like the huge Asian "Brown Cloud" that hangs over megacities in China to the carbon dioxide emissions accumulating in the atmosphere from millions of individual sources, humans are creating a new physical reality.

Since the beginning of industrialization humanity has been running a gigantic geophysical experiment. People have been mining coal and petroleum from earth's crust, burning them and dispersing the resultant carbon dioxide into the atmosphere for some time. Carbon is also released into the atmosphere when forests burn down or wetlands dry out. According to estimates by the Potsdam Institute, Oxford University and the World Resources Institute, an additional 2,110 billion tons of carbon dioxide went into circulation between 1800 and 2014 as a result of human activities.[60] This already represents a significant disturbance of the earth's

---

58. See personal website of the artist www.david-thomas-smith.com.
59. International Energy Agency, *World Energy Outlook 2012*, Paris: IEA, 2012.
60. International Energy Agency, ibid.

carbon cycle.[61, 62] Despite global efforts, carbon dioxide emissions are still increasing; for energy consumption alone, emissions stood at 34.5 billion tons in 2012 and 36 billion tons in 2013.[63, 64]

If current trends continue, then from the beginning of industrialization to approximately 2025, the same quantity of additional carbon, in the form of carbon dioxide, will have been introduced into the atmosphere and oceans by humans, as is contained in all living organisms today.[65]

Around sixty per cent of this colossal quantity of material is absorbed by the oceans and vegetation for now; because carbon dioxide dissolves in water and plants can absorb it and convert it into biomass. But that does not mean that the element has disappeared. In the sea, acidification has already started to take place because carbon and water combine to form carbonic acid, as everyone familiar with chemistry or fizzy drinks well knows.[66]

If you've ever downed a soft drink with too much carbonic acid, you get an idea of how countless sea organisms must feel as they are disturbed by carbon dioxide. Carbon dioxide has a slightly corrosive effect; organisms partly composed of calcite—like plankton or coral reefs—are extremely vulnerable to this. Coral reefs and the shells of organisms like diatoms, which are at the beginning of the food chain, are therefore endangered by acidification, which leads to the coral bleaching. While there are methods by which coral reefs buffer acidity, this is only possible to a certain extent.[67]

61. For a graphic representation of CO_2 emissions, see the Oxford University project at http://trillionthtonne.org.

62. UNESCO and UNEP, *The Global Carbon Cycle*, Paris: UNESCO, November 2009.

63. PBL Netherlands Environmental Assessment Agency, *Trends in global CO2 emissions: 2013 Report*, Den Haag, 2013.

64. Global Carbon Project, *International Geosphere-Biosphere Programme*, November 19, 2013, http://www.igbp.net/news/news/news/annualglobalcarbonemissionssettoreach-record36billiontonnesin2013.5.30566fc6142425d6c91195a.html.

65. Dr. Thomas Dittmar, Max-Planck-Institute for marine Microbiology, personal communication, August 2011.

66. Jeremy Jackson, "Ecological extinction and evolution in the brave new ocean," *Proceedings of the National Academy of Sciences*, vol. 105, suppl. 1, August 12 2008: 11458–465.

67. Andreas J. Andersson et al., "Partial offsets in ocean acidification from changing coral reef biogeochemistry," *Nature Climate Change*, published online, November 17, 2013, http://www.nature.com/nclimate/journal/vaop/ncurrent/full/nclimate2050.html.

The negative consequences of acidification are considerable because of how they affect the marine food chain.[68]

This development worries me much more than higher temperatures on land. The oceans cannot absorb unlimited quantities of carbon dioxide and, if a tipping point is ever reached, warmed seas could change from being repositories of greenhouse gases into becoming sources of greenhouse gas emissions. If the seas become warmed to their depths, and there are already indications that this is happening, frozen methane gas could thaw and be released into the atmosphere. Methane is a much more potent heat-trapping gas than carbon dioxide.[69, 70]

At the beginning of industrialization, the concentration of carbon dioxide in the atmosphere was around 280 parts per million molecules of air (ppm). At the time of the UN Climate Summit in Rio de Janiero, Brazil, in 1992, it had reached 356 ppm. In May 2013, at the monitoring station on the summit of the Mauna Loa volcano on the Big Island of Hawaii, the value exceeded 400 ppm, and according to current trends it will reach 440 ppm by 2040—a threshold that climate researchers regard as critical.[71]

Scientists from the Intergovernmental Panel on Climate Change (IPCC) continue to agree that emission of greenhouse gases caused by humans will increase the earth's average temperature by at least two degrees Celsius by the end of this century. Apparent breaks in global warming trends appear to have more to do with complicated feedback mechanisms between the atmosphere and the oceans and a lack of measuring stations in the Arctic, rather than with incorrect scientific assumptions about climate change.[72, 73]

68. The *Seattle Times* together with the Pulitzer Center on Crisis Reporting published an excellent yet alarming report on ocean acidification entitled "Sea Change", September 12, 2013, http://apps.seattletimes.com/reports/sea-change/2013/sep/11/pacific-ocean-perilous-turn-overview/.

69. Timothy Lenton, Hans Joachim Schellnhuber et al., "Tipping elements in the Earth's climate system," *Proceedings of the National Academy of Sciences*, vol. 105, no. 6, (February 12, 2008): 1786–1793.

70. For an overview of the ocean in the Anthropocene see Davor Vidas, "The Anthropocene and the International Law of the Sea," *Philosophical Transactions of the Royal Society* –A, vol. 369 (2011): 909–925.

71. For continual measured data, see http://www.esrl.noaa.gov/gmd/ccgg/trends/.

72. IPCC, *Climate Change 2013: The Physical Science Basis*, Geneva, Switzerland, 2013.

73. Kevon Cowtan and Robert G. Way, "Coverage bias in the HadCRUT4 temperature series and its impact on recent temperature trends," *Quarterly Journal of the Royal*

Two degrees Celsius does not sound like much, which is true when it refers to normal daytime temperatures. However, when applied to the *global average temperature of the earth*, it is comparable to human body temperature, where an increase of two degrees can make the difference between normal well-being and life threatening illness. In extreme scenarios, average temperature could rise six degrees by the end of the century, and even higher in some regions. At the moment, the global average temperature is about five degrees warmer than at the peak of the last Ice Age, when glaciers in the Northern hemisphere soared hundreds of feet high. An increase in the average global temperature of five or six degrees would portend the beginning of a "Heat Age."

There is no certainty about what all this additional carbon dioxide will do to the earth. Scientific models are imprecise and not all future changes can be predicted. But, does that give us the liberty to play down the impact that people have on the climate the way that interest groups do, especially in the United States? Believing the climate change skeptics is taking an enormous risk for, if they are wrong, we will face a dangerous, perhaps irreversible situation. If the critics are proven right, little change will occur, except perhaps reasonable investments in environmental protection and renewable energy sources.

Underpinning the arguments of climate change skeptics is an assumption that humans are only a small factor in world events, so negligible that they cannot possibly trigger serious consequences. We shall simply carry on, they say. Advocates of this attitude think little of the Anthropocene idea. However, many scientific findings undermine the argument that human actions are only a trivial factor in earth events. The magnitude of the human factor is shown by one US Geological Survey study, according to which humans emit 135 times more carbon dioxide than all volcanoes combined.[74] Axel Kleidon from the Max Planck Institute for Biogeochemistry, in Jena, Germany, states that annual human consumption of free energy stands at approximately 50 terawatts, mainly due to burning fossil fuels and cultivating crops. This is equivalent to between five and ten

Meteorological Society, October, 2013, http://onlinelibrary.wiley.com/doi/10.1002/qj.2297/abstract.

74. Terry Gerlach, "Volcanic versus anthropogenic carbon dioxide," *EOS, Transactions of the American Geophysical Union,* vol. 92, no. 24, (14 June 2011): 201–208.

per cent of free energy available. According to Kleidon, this is significantly more free energy than is produced by all the volcanoes, earthquakes and other tectonic events, combined.[75]

Population growth, and space and energy requirements are the powerful forces propelling us out of the Holocene into the Anthropocene. Our abilities and needs, our knowledge and emotions, are beginning to transform not only the surface of the earth but also the future course of evolution—this is the fourth dimension of the Anthropocene. We are running populations of many plant and animal species down to the point of extinction. Biologists talk about a sixth wave of extinction in the history of the earth that is now underway, due to cutting down tropical rainforests, overfishing, overhunting, and a general loss of habitats.[76] Geologists will see a reflection of this in the fossils that will be left from our epoch.[77]

Further, humans are beginning to create new life-forms through interbreeding, gene technology and more recently, biotechnical design. Life-forms of the future might be products of the human imagination: a scientist's bracing walk through the forest might spawn a new form of life some months later. Trading and transport routes are bringing about large-scale changes in the distribution of animals, plants and other organisms and may determine whether they continue to exist at all. The figure that probably best symbolizes the transition from the Holocene to the Anthropocene is how matter is distributed among life-forms. According to an estimate by Vaclav Smil, 10,000 years ago, humans and their livestock were a mere 0.1 per cent of the entire live weight of mammals. The other 99.9 per cent was being used by elephants, deer, gorillas, and so on. According to Smil's estimate, 90 per cent of today's mammalian matter is part of the soon-to-be eight to ten billion people on the earth, along with their

75. A. Kleidon, "How does the earth system generate and maintain thermodynamic disequilibrium and what does it imply for the future of the planet?," contribution to Theme Issue "Influence of Nonlinearity and Randomness in Climate Prediction," *Philosophical Transactions of the Royal Society A*, http://arxiv.org/pdf/1103.2014v2.pdf.

76. Anthony D. Barnosky et al., "Has the Earth's sixth mass extinction already arrived?" *Nature*, (March 3, 2011) vol. 471, no. 51–57.

77. See also the outstanding books by Edward O. Wilson and Jean-Christophe Vié et al., *Wildlife in a Changing World—An Analysis of the 2008 IUCN Red List of Threatened Species*, Gland: IUCN 2009 and Arthur D. Chapman, *Numbers of living species in Australia and the world*, Canberra: Australia Biodiversity Information Service, 2009.

billions of cattle, pigs, dogs and other domesticated creatures.[78] Human influence on the current and future course of evolution has become huge.

In the course of my work as an environmental and science journalist over the past few years, I have experienced at first hand many of the problematic phenomena that scientists believe imply the end of the Holocene. I have stood in Borneo and in Amazonia, in the middle of a blazing rainforest. I have been scuba diving off the coasts of Mexico and Indonesia, observing devastated coral reefs. I have witnessed the clearing of old-growth forests on Vancouver Island in British Columbia, Canada and in Finland. I have traveled miles below the earth's surface to places where nuclear waste is supposed to be stored for millions of years. I have trekked across melting glaciers in the Alps and have directly experienced the fragility of ecosystems in the Himalayas. In New Zealand and Central Africa, I have observed some of the rarest animal species in the world. In laboratories in the US and Europe, I have explored how biotechnologists are starting to control the forces of life. Starting with the first-ever UN climate summit held in Berlin in 1995, I have reported from many global conferences on protection of the climate and biodiversity.

After all these experiences, it seems quite obvious to me that humans have become a geological factor during the Holocene. I saw many terrible things during the course of my investigations and met people whose livelihoods had been stolen by rainforest clearing. Some experiences made me wonder whether we are witnessing the collapse of our civilization, a global variation on what Jared Diamond so vividly described of past civilizations all over the earth.[79] So much has been destroyed, so much is vanishing. But what does this really mean? The end of the Holocene is, at the same time, a beginning. What I felt was a fusion of nature, of people and technology into something new.

When I began to think about the Anthropocene idea, I realized that this fusion process touches our modern Western worldview to the core.

---

78. Vaclav Smil, *The Earth's Biosphere: Evolution, Dynamics, and Change*, Cambridge, MA: MIT Press, 2002, quoted in Gaia Vince, "A Global Perspective on the Anthropocene," *Science*, (7 October 2011) 32-37.

79. Jared Diamond, *Collapse: How Societies Choose to Fail or Succeed*, New York: Penguin, 2011.

Humans are accustomed to neatly categorizing "people" and "environ-ments," "nature" and "culture," "economy" and "ecology," "geology," and "technology." It is on such distinctions that Western society has been based. In the Holocene, there was always a "big world out there," the "great outdoors," an infinite natural world that seemed inexhaustible, at least as late as the 1950s, even to the environmental visionary Rachel Carson.[80] But, in the Anthropocene there is only "the great inside." jointly shaped by each one of us in everyday life, like global interior designers. We are not separate from our environment.

To understand the extent to which we human beings are changing the earth, you do not have to live in an urban region in China with a hundred million neighbors, or on the agricultural plains of the American Midwest that stretch to the horizon, nor on the edge of a burning rainforest. Today, it is enough to stop for a moment and realize that with every meal, we alter distant ecosystems as if by remote control because the ingredients come mostly from different continents or even ecological hotspots: palm oil grown in former rainforests or industrially produced pork. Just by get-ting into a car, turning on the heat or air conditioning, or going on vaca-tion by plane, we impact the world's climate. Each time we reach for our Smartphones, we are holding to our ear an assortment of rare metals that have come from dozens of different mines around the world!

In the Holocene, the world seemed boundless. Now everything we do rebounds on us. In the future it will be difficult, if not impossible, to cling to traditional demarcations and make distinctions between "natural" events and man-made phenomena. Has the beautiful plant growing at the side of the road been cultivated in a bio-lab or is it wild? Are cranes, now rare, still wild or already domesticated because they feed on genetically modi-fied corn? Is that an ancient coral reef or a new one that's grown up around a shipwreck? Do the clouds in the summer sky come from Mother Nature or are they jet plane vapor trails? Is that an old-fashioned thunderstorm

80. Callum Roberts, *The Unnatural History of the Sea*, Washington DC: Shearwater, 2009.

brewing overhead or one that wouldn't be there if not for climate change? Our descendants may not even ask themselves these kinds of questions. Storms and floods in the future cannot be called natural disasters; they will be "cultural calamities." An initiative by US environmental conservationists to name future hurricanes after politicians who have not acted to prevent climate change seems logical.[81]

Conversely, "natural wonders" in the future are more likely to be "wonders of civilization"—biologically rich landscapes or blossoms of anthropogenic evolution.

Even the most ardent advocates of the Anthropocene idea would never claim that human activity is completely replacing nature. Essentially, what is happening is that humans are becoming *the dominant force of change* on earth. Two prominent geologists, Charles H. Langmuir from Harvard University and Wally Broeker from Columbia University express what is taking place, in the following way: "The rise of human civilization is a transformative event in planetary history. For the first time a single species dominates the entire surface, sits at the top of all terrestrial and oceanic food chains, and has taken over much of the biosphere for its own purposes."[82]

We humans have grown up as children of the Holocene but a new phase is clearly arising, within the lifetime of our species.[83] This phase is characterized not only by measurable environmental changes but also by transitions in consciousness, learning, connectedness, cooperation and other capabilities with positive potential.

As our actions become more global, so does our environmental awareness. The more materials and living things we set in motion, bringing them to new places in new combinations, the more we expand our repertoire to track and influence these changes. And the more our actions reach toward the future, the greater our capacity grows to develop scientific models of change: from making projections about the world climate in fifty years' time to events like the intergalactic merger of the Milky Way

---

81. See www.climatenamechange.org.

82. Charles H. Langmuir and Wally Broecker, *How to build a habitable planet—The Story of Earth from the Big Bang to Humankind,* Princeton University Press, 2012.

83. Andrew Revkin has compared the developmental stage of our species with puberty: http://dotearth.blogs.nytimes.com/2011/09/20/maturing-teens-maturing-species/.

with the Andromeda Nebula due 3.7 billion years from now.[84] The very fact that we can classify our collective actions on the scale of Earth's history is in itself a positive sign. Departure from the Holocene may be happening under some very frightening circumstances, but the Anthropocene is not destined to be a scary thing.

84. See "NASA's Hubble Shows Milky Way is Destined for Head-On Collision," May 31, 2012, http://www.nasa.gov/mission_pages/hubble/science/milky-way-collide.html.

FOUR Signals of Earth Time

A S YOU READ THESE LINES, various time scales are at work. It takes
just milliseconds for your brain to interpret these letters by means
of synapses and neurons, and mere seconds for you to put what I
am writing into context. It has taken months, years and even decades for
your brain to mature to its current state. Your brain has a history going
back hundreds of thousands or even millions of years. The molecules in
your brain that perform these feats are billions of years old.

Thanks to the work of brain researchers, molecular biologists, evolutionary scientists and geologists, we have become familiar with the different time scales in which our lives exist—from the femtoseconds of the
quantum world to billions of years of cosmic history.

Contemporary science makes it astonishingly easy to go on a mental
time journey, taking enormous leaps through long periods. But it wasn't
that long ago that the intellectual elites of Europe or America had a completely repressed relationship with such timeframes. A mere two to three
hundred years ago it was considered heretical, and a sure path to damna-
tion, to believe that the earth was any older than 6,000 years. The French
natural scientist Georges-Louis Leclerc (later Comte de Buffon), was one
of the first, in the late 18th century, who dared question the timeframe that
church officials had deduced from the Old Testament. By examining the
rate at which molten iron cooled, Buffon put forward the hypothesis that
the earth, with an inner core of iron, had to be about 75,000 years old.[85]

85. Georges Louis LeClerc Comte de Buffon, Des époques de la nature, in: "Histoire
naturelle, générale et particulière." Supplement Vol. V, Paris, Imprimeries royale, 1778. For
a general history of time computation, see also Daniel Rosenberg and Anthony Grafton,
Cartographies of Time—A history of the Timeline, Princeton Architectural Press, 2010.

Even this overly conservative calculation got Buffon into trouble with his university and with the Catholic Church. To this day, creationists, mostly in the United States, use absurd claims to try to prove that the earth is only a few thousand years old.[86]

Time has been understood differently in non-Western cultures, which have been more open to the idea of a deep past. In the Hindu religion a single day and night in the life of the creator god Brahma lasts 8.64 billion years and a year, for Brahma, lasts 3.11 trillion years. The Buddha described how an enormous mountain could be worn away by rubbing it with a silk scarf before one world cycle, or *Maha-Kalpa*, had passed.

In Western societies, it has taken much longer than in the East to comprehend the temporal dimensions of existence. When, in the nineteenth century, scientists began to measure the age of rocks and chemical compounds using sophisticated instruments, an amazing expansion of time began, from the theologians' notion of a 6,000 year existence to the realization that the earth may actually be 4.57 billion years old.

What were humans to do with this extended past? Awestruck by the magnitude of time's expansion, a group of natural scientists formed an exclusive assembly of "terrestrial timekeepers," holding the first International Geological Congress in Paris in 1878. Since the foundation of the International Union of Geological Sciences (IUGS), the self-appointed task of its members has been to divide the past into logical intervals and give names to the various epochs. The chief concern was to tell the "longest story" but in reverse—the story of the earth itself—dividing it into compelling chapters. Scientists who do this are called stratigraphers, after the strata or rock layers that lie beneath us. They spend their professional lives at locations where strata have broken up and become visible on the surface, or where core drilling has revealed earth's crust as a multi-layered cake of various colors, thicknesses and compositions. They use complex machinery to determine the age of the stones and minerals they find and generally, the deeper the strata, the older they are.

The largest and most important constituent sscientific body in the IUGS organization of terrestrial timekeepers is the International Com-

---

86. My favorite example of these absurdities is the website http://creation.com/age-of -the-earth. Creationists argue that Earth is extremely young while showing a picture of Venus.

PERIOD	EPOCH	ERA
QUATENARY	"ANTHROPOCENE"	
	HOLOCENE	
	PLEISTOCENE	
TERTIARY	PLIOCENE	CENOZOIC
	MIOCENE	Present to 66 million years ago
	OLIGOCENE	
	EOCENE	
	PALEOCENE	
CRETACEOUS		
JURASSIC		MESOZOIC
TRIASSIC		66 to 252 million years ago
PERMIAN		
CARBONIFEROUS		
DEVONIAN		
SILURIAN		
ORDOVICIAN		PALEOZOIC
CAMBRIAN		542 to 252 million years ago

mission on Stratigraphy. In its numerous task groups and committees, scientists discuss which "signals" in the rock layers justify separating (and naming) one sub-plot of earth's history as distinguished from another. This is done in a most precise, scientific manner in order to create a reliable classification and avoid having to alter the boundaries of these epochs every few years. Only in the naming is some degree of freedom and creativity permitted. Inspiration has been found in a Celtic tribe (the Ordovician), or in the idyllic English landscape of Devonshire (the Devonian), names being determined by where researchers found rocks pertaining to a certain period.

The time dimensions with which stratigraphers deal are formidable, especially compared to a non-geologist like me, who works on the nanoscale of daily life's rounds of meetings, deadlines, invitations, children's birthdays, and is delighted to have a handle on next week. Presumably, stratigraphers also have to deal with such mundane things but somehow they manage to live in two temporal zones at the same time. And so, over the past decades, they have created an impressive color-coded chart to show the geological eras, the names of which are stacked on top of each other much like the layers of rock.[87]

This chart depicts great eons like the Archean, the phase 1.5 billion years ago marked by the advent of bacteria, immediately after the creation of life; it depicts the Cenozoic or "new animal" Era, which encompasses the entire 66-million-year rise of mammals from the dinosaurs' extinction to the present day. It shows periods such as the Jurassic (that has become better known since the film *Jurassic Park*), and vast epochs covering many millions of years like the Pleistocene or the current Holocene, a smaller unit on the geological timescale. Even this smallest scale is beyond normal human imagination.

So, this is another reason why Paul Crutzen's declaration of the Anthropocene, at that conference in Mexico back in February 2000, was such a huge statement. What he did was tantamount to driving humanity out of its ancestral geological home in the Holocene, and resettling it in new chronological territory, the Anthropocene. He effectively remapped the various timescales in which our existence is recorded—from the

87. International Chronostratigraphic Chart, see www.stratigraphy.org.

nanoseconds of stock exchanges to the four-year rhythm of politics—to much longer geological timescales. In doing so, he enabled human history to become the subject of geological examination. Human history became a part of deep earth history, an area that had previously been almost exclusively the realm of biologists and geologists.

After the conference in Mexico, it quickly became apparent that there were hundreds and thousands of extant observations, studies and analyses showing that modern humans were indeed changing the Earth in a radical, long-term manner within a very short space of time, so much so, that future geologists will notice these changes. Scientists had been gathering evidence of the traces left behind by humanity for quite a while. Examples of these include artificially created elements, radioactive fallout from atom bomb tests, an increase in atmospheric carbon dioxide, plastic waste, and the colorful assortment of archaeological substrata beneath cities.

What had been missing up until then was a term to summarize these changes and someone sufficiently prominent to make such a term popular. To his surprise, after the conference, Paul Crutzen found out that another scientist, Eugene F. Stoermer, a limnologist at the University of Michigan, had already used the term Anthropocene back in the 1980s. In a book written in 1992, journalist Andrew Revkin of the *New York Times* claimed: "We are entering an age that might someday be referred to as, say, the Anthrocene."[88] Andrew certainly earns a warm round of applause for almost nailing the magical new word.

In 2000, Crutzen contacted Stoermer, as is proper when two scientists have arrived at the same conclusions independently of one another. He suggested that they publish the Anthropocene idea together. Stoermer agreed, later saying: "I began using the word 'Anthropocene' in the 1980s, but I never formalized it until Paul contacted me."[89]

Starting with a short article in the newsletter of the International Geosphere-Biosphere Programme (IGBP), the Anthropocene idea was

88.   Andrew Revkin, *Global Warming: Understanding the Forecast*, Abbeville Press, 1992, quoted in Will Steffen et al. "The Anthropocene: conceptual and historical perspectives," *Phil. Trans. R. Soc.*, A 13, vol. 369, no. 1938 (March 2011): 842-86.

89.   Quoted from J. Grinevald, *La Biosphère de l'Anthropocène: climat et pétrole, la double menace*, Editions Médecine and Hygiène, Geneva, Switzerland, Repères transdisciplinaires, 2007: 243.

born.[90] Two years later Crutzen called for the renaming of the Holocene to Anthropocene, in an article entitled "Geology of Mankind" published in the influential scientific journal, *Nature*. "Unless there is a global catastrophe—a meteorite impact, a world war or a pandemic—mankind will remain a major environmental force for many millennia."[91]

Together with the renowned environmental researcher Will Steffen and historian John R. McNeill, Crutzen clarified his idea in 2007, and made a suggestion as to when the start of the Anthropocene might have been.[92] According to them, the "pre-phase" of the Anthropocene ran from the time of the first human-made fire to the first fire inside a steam engine. The Anthropocene really began in 1800 because that is when scientific enlightenment and technology-driven industrialization produced measurable geological, chemical, and biological changes on earth.

This pioneering article by Steffens, McNeill, and Crutzen contained an illustration that attracted a lot of attention: it showed various developments since 1945, from world population and energy consumption, to international tourism and the number of McDonald's restaurants. All parameters showed a steep increase. The illustration captured in a nutshell what Crutzen and his colleagues call the "Great Acceleration." This process, they wrote, "took place in an intellectual, cultural, political and legal context in which the growing impacts upon the Earth System counted for very little in the calculations and decisions made in the world's ministries, boardrooms, laboratories, farmhouses, village huts, and for that matter, bedrooms."

Since the first papers describing it, the Anthropocene idea itself has undergone a "Great Acceleration," as shown by the increases below on one search engine:

In 2003, there were 413 Google hits for the term Anthropocene;

In 2011, there were 450,000;

90. Paul J. Crutzen und Eugene F. Stoermer, *IGBP Newsletter 41*, (May 2000), http://www.igbp.net/news/opinion/opinion/haveweenteredtheanthropocene.5.d8b43c12bf3be638a8000578.html.

91. Paul J. Crutzen, "Geology of Mankind," *Nature*, vol. 415, no. 23, (2002).

92. Will Steffen, Paul J. Crutzen und John R. McNeill, "The Anthropocene: Are Humans Now Overwhelming the Great Forces of Nature?," *Ambio*, vol. 36, no. 8, (December 2000): 614-621.

In 2013, there were 1,070,000.[93]

Obviously, Google hits do not say anything about how correct or profound an idea is. This can be illustrated perfectly by the number of hits returned by a search for "Glenn Beck." For a term as esoteric as "Anthropocene," this increase is considerable.

The Anthropocene idea touches a nerve with roots deep in the history of our civilization. Critical debates on humankind's role on earth were taking place as long ago as the sixteenth century. Christian Europe, after China, the second center of science and technology, had begun to conquer the known world. During this period, a French artist named Jean de Gourmont created a symbol that characterizes this process. De Gourmont drew the most up-to-date world map of his time in the form of a human face. Rather than give this face the appearance of a king, he depicted it as a fool with a double-peaked, bell-tipped cap and jester's staff.[94]

Known as "The Fool's Cap," the drawing was produced circa 1575, in the heyday of early explorers and European colonizers. It was the time of a worldwide race for raw materials and land estates, finding silver mines in Peru and spice plants on the Maluku Islands (Moluccas). Meanwhile, British miners were already excavating two hundred thousand tons of coal a year. Emperor Charles V had broken the Christian Church's prohibition on limiting interest, thus enabling Dutch merchants to lend him money, and unleashing a wave of speculative investment that has lasted to the present day.

De Gourmont's fool's head was regarded as a warning against earthly fantasies of power, pride and folly. Tattooed on the fool's forehead is a proclamation of how absurd it is for men to fight over the earth with sword and flame. De Gourmont calls the earth the "world point." This was a revolutionary term at a time when there was no telephone, no Internet, when no one spoke of a "global village" and when it was risky to question the

93. James Syvitski, Anthropocene, "An epoch of our making," *Global Change*, issue 78, (March 2012): 14. The figures are the number of hits displayed by my own computer, on 21 November 2013.

94. Richard Helgerson, "The Folly of Maps and Modernity, Literature, Mapping and the Politics of Space" in *Early Modern Britain*, Andrew Gordon and Bernhard Klein (eds.), Cambridge University Press, 2001:241-262.

biblical creation story. The Frenchman was thus one of the first to real-ize that the world is truly small, rather than the infinite space it seemed to conquerors at the time.

On the jester's scepter is written the "infinite vanity of human beings." Above the fool's head hangs the command: "Know Thyself." For a long time, the map was filed in the humor section of map collections. But it is one of the most sharp-witted, serious reflections on the relationship of humanity with the Earth.

Many others shared this critical line of thinking, including Hans Carl von Carlowitz, an early eighteenth century tax accountant and mining administrator at the royal court of Freiberg in Saxony. He determined how mining had led to a decline in forests across the whole of Europe. He lam-basted the wastage of wood by the rich and in 1713 introduced the concept of sustainability.[95]

At the beginning of the nineteenth century, another person who recog-nized the signs of the times, early on, was the German polymath, Alexan-der von Humboldt. After his voyages of discovery to South America and Russia, during which he explored the riches of the natural world, he wrote his major work *Cosmos*. He described the earth as a "world organism" and claimed that nature first has to be profoundly understood before we can use it a meaningful way.[96] Humboldt was one of the most important pio-neers of the Anthropocene idea but he was far from using the term itself. There was no serious geological classification at that time, and Buffon had recently taken criticism for his estimate that the world was 75,000 years old. It was not until later in the nineteenth century that it became possible to propose geological timescales.

The Italian geologist Antonio Stoppani came very close to the term when he spoke of an "anthropozoic era" and referred to humanity as a "new telluric force, which in power and universality may be compared to

95. Carl von Carlowitz, *Sylvicultura Oeconomica oder haußwirthliche Nachricht und Naturmäßige Anweisung zur Wilden Baum-Zucht*, second edition reprint, Leipzig: heirs of the late Johann Friedrich Braun, 1732, Remagen-Oberwinter: Verlag Kessel, 2009.
96. Alexander von Humboldt, *Kosmos—Entwurf einer physischen Weltbeschreibung*, (Cosmos—a General Survey of Physical Phenomena of the Universe) second volume, Stuttgart and Tübingen: F. G. Cotta'scher Verlag, 1847.

the greater forces of Earth."[97] In 1864, George Perkins Marsh published his far-sighted book, *Man and Nature*, in which he wrote a detailed description of human influences on the environment, and ended with a public appeal: "The collection of phenomena must precede the analysis of them, and every new fact, illustrative of the action and reaction between humanity and the material around it, is another step toward the determination of the great question, whether man is of nature or above her."[98]

In the early twentieth century, many thinkers set out to make the interplay between man and nature comprehensible. An essay was published in 1915 in the *Zeitschrift der Deutschen Geologischen Gesellschaft* (Journal of the German Geological Society) by a young scientist named Ernst Fischer with the title "Der Mensch als geologischer Faktor" (Man as a geological factor). Listing numerous environmental changes, Fischer concluded with a warning that the greater humanity's impact, the greater the danger of destruction, with the provision that "increased intellectual activity" and "reflected self-knowledge" could balance this out. Unfortunately, Fischer died young in World War I before his essay's publication.[99]

Also worthy of mention are R.L. Sherwood's "Man as a geological agent" and Erwin Fels's book "Der Mensch als Gestalter der Erde" (Man as a transformer of Earth).[100, 101]

Other scientists like the Russian geochemist Vladimir Ivanovich Vernadsky and his teacher A.P. Pavlov, the French scientists Édouard Le Roy and the Jesuit Pierre Teilhard de Chardin became more famous for their interpretations. They theorized about the growing impact of human consciousness and cognition on the environment and on evolution.[102] Pavlov, a geologist, even spoke of an "anthropogenic era."

97. Antonio Stoppani, Corso di Geologia, Verlag: G. Bernardoni, E.G. Brigola Editori 1871-1873, Milano, 1871, quoted in W.C. Clarke et al., "Sustainable Development of the Biosphere", *Environment*, vol. 29, issue 9, (1987): 25-27.

98. George P. Marsh, *Man and Nature*, New York, C. Scribner, 1864, quoted from the edition edited by David Lowenthal, Harvard University Press, 1965.

99. Ernst Fischer, "Der Mensch als geologischer Faktor," Zeitschrift der Deutschen Geologischen Gesellschaft, vol. 67, (1915): 106-149.

100. R.L. Sherwood, *Man as a geological agent—an account of his action on inanimate nature*, London: H.F. & G. Witherby, 1922.

101. Edwin Fels, *Der Mensch als Gestalter der Erde*, Bibliographisches Institut AG in Leipzig, Leipzig, 1935.

102. Vladimir Vernadsky, *Geochemistry and the Biosphere*, Synergetic Press, Santa Fe,

Vernadsky suggested that human influence was affecting all the elements, changing geochemical cycles and the thermodynamic balance of the biosphere. He worked out how closely interwoven the biosphere and humanity are: "In the 20th century, for the first time in history, Man knew and embraced the biosphere, completed the geographical map of the Earth, and settled all over its surface...Mankind taken as a whole is becoming a powerful geological force. Humanity's mind and work face the problem of reconstructing the biosphere in the interests of freely thinking mankind as a single entity." Vernadsky popularized the use of the term "noosphere" to describe a world indelibly marked by human thought and cognition, in which the geosphere and biosphere would be changed by humanity. Stemming from this, Teilhard de Chardin wrote *The Phenomenon of Man*, in the 1930s, although it was not published until 1955, posthumously. It was a new Christian-based philosophy about the fusion of man and nature, of consciousness and materiality. It found little favor with the Vatican.

During the 1960s and 1970s, people were so focused on the short-term risk of nuclear annihilation that long-term ideas were not much in vogue. In the 1980s, German biologist Hubert Markl, later to become president of the Max Planck Institute, diagnosed the beginning of an "Anthropozoikum" in his book *Natur als Kulturaufgabe* (Nature as a cultural task) when he wrote: "Whether he [mankind] masters or fails in this task, he will have to take responsibility for it. There will be no difficulties dating this latest mass extinction of fauna: it's happening here and now."[103]

The long history of precursors of the Anthropocene concept begs the question as to why it took until Paul Crutzen's famous statement for the idea to gather any momentum. I think there are several reasons for this: Firstly, the time must be ripe for an idea. In 2000, recognition of anthropogenic climate change and the sense of global networks like the Internet or international trade were certainly factors. Secondly, it is helpful when the advocate of an idea already has international standing. Crutzen's status

NM, 2007; Vladimir Vernadsky, *La Biosphere*, Librairie Felix Alcan, Nouvelle collection scientifique, Paris, 1929; Édouard Le Roy, *Les origines humaines et l'evolution de l'intelligence*, Paris, 1928; Teilhard de Chardin, "Hominization," (1923), and Teilhard de Chardin, *The Phenomenon of Man*, Paris, 1955. The latter is available online at http://arthursbookshelf.com /other-stuff/phenom10.html.

103. Hubert Markl, *Natur als Kulturaufgabe—Über die Beziehung des Menschen zur lebendigen Natur*, DVA, 1986.

as "most-cited scientist" brought him considerable attention. Thirdly, the medium of communication also plays a role. Crutzen did not publish a book; instead, he presented his hypothesis at a scientific conference and in an article in *Nature*, one of the most prestigious scientific journals in the world.

It was due to Crutzen's postulation being a genuinely formulated scientific hypothesis that his did not remain just one idea among many. It has since undergone rigorous examination by other scientists, which could lead to its eventual official recognition by the keepers of earth time.

In 2009, a thirty-member scientific task group was formed under the auspices of the ICS (to be called the Anthropocene Working Group of the Subcommission on Quaternary Stratigraphy), led by the geologist Jan Zalasiewicz, from the University of Leicester. Its volunteer members collect evidence worldwide and examine whether the renaming of our epoch is justified or useful.[104] The Working Group plans to publish its findings in a book, to be issued in 2016.

The timekeepers have set up hard and fast rules to classify the Earth's 4.6-billion-year history. The past is to be divided according to uniform criteria, and this is accompanied by a protracted examination process. Moving the boundaries between epochs or eras may take decades.

In February 2011, the Anthropocene Working Group published its first interim results, in a special edition of the *Philosophical Transactions of the Royal Society*, which is the oldest scientific journal in the world.[105] Only a few months later, a group of experts met at the Geological Society in London to lay out the scientific indicators that favor the Anthropocene idea. Shortly after the conference in London, an editorial in *Nature* approved the renaming of the epoch: "Official recognition of the Anthropocene would focus minds on the challenges to come."[106] The French newspaper *Le Monde* and the British *Economist* published cover stories headlined:

---

104. See Jan Zalasiewicz et al., "Are we now living in the Anthropocene?," *Geological Society of America Today*, vol. 18, no. 2, (2007): 4-8 and the task group's website: http://quaternary.stratigraphy.org/workinggroups/anthropocene/.

105. Jan Zalasiewicz et al., "Stratigraphy of the Anthropocene," *Philosophical Transactions of the Royal Society of London*, A 369 (2011): 1036-1055.

106. Nature magazine editorial board, "The human epoch," *Nature*, vol. 473, (19 May 2011): 254, see http://www.nature.com/nature/journal/v473/n7347/full/473254a.html.

"Welcome to the Anthropocene."[107] In 2011, the Geological Society of America named its annual conference "From the Archean to the Anthropocene," as if it were self-evident. American geologist, Susan Trumbore, of the Max Planck Institute, thinks that any controversy is irrelevant. She says: "The Anthropocene is an obvious reality; we are leaving our traces almost everywhere."[108] Stanley Finney, Chairman of the International Commission of Stratigraphy, expressed favor for the Anthropocene idea for the first time in the autumn of 2013.[109]

There are already lively debates about the official start date of the Anthropocene, should it be formally declared, because if a new geological epoch is introduced into a series of earlier epochs, there must be exact boundaries set that accord to geological rules. This is how each geological era has been defined, using distinct markers that are found around the world, known as "Global Boundary Stratotype Sections and Points" (GSSP). For example, the boundary between the Holocene and the Pleistocene is demarcated by a drill core sample taken from Greenland, showing the intense warming that occurred just after the last Ice Age. The Cambrian era, which began 541 million years ago, is delineated using fossils from a specific sea organism, the *Treptichnus pedum*, and from similar first representatives of multicellular, highly organized life-forms. The GSSP that delineates that particular boundary is the collection of fossils found at Fortune Head, in Newfoundland, Canada.

So how and when could the Anthropocene be demarcated from the Holocene? Stanley Finney, at the ICS, warns that extreme scientific precision has to preside. He has a long list of questions that have to be worked through in order to ratify a new epoch.[110] Among these is the task of finding a characteristic marker that represents the shift from Holocene to Anthropocene.[111]

107. "Welcome to the Anthropocene," *The Economist*, 26 May 2011.

108. Axel Bojanowski and Christian Schwägerl, "Debatte um neues Erdzeitalter: Was vom Menschen übrigbleibt," (Debate on a new geological epoch: the traces left by man) Spiegel Online, 4 July 2011.

109. Stanley C. Finney, The 'Anthropocene' as a ratified unit in the ICS International Chronostratigraphic Chart: fundamental issues that must be addressed by the Task Group, *Geological Society, London, Special Publications,* first published October 24, 2013; doi 10.1144/SP395.9.

110. S. C. Finney, "The 'Anthropocene' as a ratified unit in the ICS International Chronostratigraphic Chart: op. cit., see footnote 102.

111. David Biello, "How long have humans dominated the planet?," *Scientific American*

Steffen and Crutzen have suggested the beginning of the Industrial Revolution, at the end of the eighteenth century, as a possible date because that is when massive amounts of carbon dioxide were released into the atmosphere and oceans. A solid representation of this would be a core sample that shows the increasing levels of carbon dioxide in the atmosphere. In 2014, the Working group proposed 1950 as a starting line for the Anthropocene. Atom bomb tests that have taken place since then have left behind a layer of radioactive isotopes spread around the world that will remain measurable for millions of years. Concurrently, the "Great Acceleration" in economic activity took place and the resulting plastics, toxins and artificially created minerals have a high potential of staying behind as "technical fossils."[112, 113] Potential markers for this starting date would be the rocks that were metamorphosed by atom bomb explosions.

Geologists Jan Zalasiewicz and Mark Williams suggest some less negative markers as possibilities, such as the emergence of megacities that they compare to the emergence of more complex life-forms at the beginning of the Cambrian period. "The record of events leading to the development of the complex trace fossils of megacities is similarly complex, finding its roots in the technology of great apes, through the stone tools of early humans, to the monumental structures of the past 9,000 years, culminating in the megacities. Each step can be seen as a staging post to the main event, in this case, taken as the trace fossil systems of the megacities."[114, 115] These scientists argue that megacities, with their underground train tunnels, building and technology debris, and landfills, will remain and become complex fossil structures, that mark the beginning of

Online, (December 6, 2013), http://www.scientificamerican.com/article.cfm?id=length-of -human-domination&WT.mc_id=SA_Facebook.

112. Jan Zalasiewicz et al., "The mineral signature of the Anthropocene in its deep-time context," 2013. In: *A Stratigraphical Basis for the Anthropocene*. Geological Society, London, Special Publications: 395. http://dx.doi.org/10.1144/SP395.2.

113. Agnieszka Galuszka, *Assessing the Anthropocene with geochemical methods*, Geological Society, London, Special Publications, first published October 24, 2013; doi 10.1144 /SP395.5.

114. Mark Williams, Jan A. Zalasiewicz et al., *Is the fossil record of complex animal behavior a stratigraphical analogue for the Anthropocene?* Geological Society, London, Special Publications, first published October 25, 2013; doi 10.1144/SP395.8.

115. M. Edgeworth, *The relationship between archaeological stratigraphy and artificial ground and its significance in the Anthropocene*, Geological Society, London, Special Publications, first published October 25, 2013; doi 10.1144/SP395.3.

the Anthropocene. Zalasiewicz also suggests markers that are *not* symbols of humanity's destructive powers but are instead symbols of human creativity and intellectual activity like gramophone records, onto which "sound" has been engraved into the material, or the points of ballpoint pens, which are made of artificial crystals.

Many archaeologists, however, make a plea to mark the beginning of the new geological epoch much earlier. In April 2013, the journal *Science* quoted numerous calls for this view at the Society for American Archaeology's 78th annual meeting: "Humans have been modifying ecosystems over a long period of time," said Bruce Smith from the National Museum of Natural History at the Smithsonian Institution in Washington DC. John Erlandson from the University of Oregon in Eugene pointed out that humans, with their extensive hunting, targeted fires and forest clearing, had already "set the stage for human domination of the earth" 60,000 years ago. Other archaeologists suggested using the advent of agriculture 11,500 years ago as the marker. Support for the view that humans had already profoundly intervened in the earth system much earlier than the onset of industrialization, came in autumn of 2013 with a study by scientists at Oregon State University. They proved that a significant increase in concentrations of methane in the atmosphere over the past 2,500 years was due solely to rice cultivation, a process that releases this potent greenhouse gas.[116, 117, 118] In their opinion, it would be advisable to simply replace the Holocene with the Anthropocene.[119]

There are certainly enough candidates for the starting date of the Anthropocene epoch. But Paul Crutzen's hypothesis throws up much more fundamental questions. Like every good idea, it has attracted criticism—and this is a good thing. It is natural for scientists to treat new

116.   Logan Mitchell et al., "Constraints on the Late Holocene Anthropogenic Contribution to the Atmospheric Methane Budget," *Science*, vol. 342, no. 6161, (22 November 2013): 964-966.

117.   Richard A. Kerr, "Humans Fueled Global Warming Millennia Ago," *Science*, vol. 342, no. 6161, (22 November 2013): 918.

118.   William F. Ruddiman, "The anthropogenic greenhouse era began thousands of years ago," *Climatic Change*, vol. 61 (2003): 261–293.

119.   Bruce D. Smith, Melinda Zeder, "The onset of the Anthropocene," *Anthropocene*, vol. 4, December 2013, online 4 June 2013, http://www.sciencedirect.com/science/article/pii/S2213305413000052.

hypotheses with skepticism and to carefully review them. The fact that the Anthropocene idea has come under fire from every direction, whether from geologists, scholars or environmentalists, is only right and proper.

Criticism by geologists is about defining a new geological epoch not only for the past but for the unknowable future. Until recently, geologists have looked almost exclusively into the past, sifting layer by layer of Earth's events. While they also spend time examining contemporary events—from the suitability of a region for oil drilling, to predictions of volcanic activity and investigation into tsunamis—the impact of humans has not played a significant role in their calculations of geological time until now, and a considerable number of geologists believe that human activity has not yet accumulated into a powerful enough signal to justify renaming the current epoch. Some even see the Anthropocene as a fanciful invention of pop culture.[120] Others do not see a globally distributed marker that qualifies as an official GSSP.[121]

"The introduction of the Anthropocene into the geological timescale would probably create more scientific problems than it solved," states Manfred Menning of the German Commission for Stratigraphy. "There is no chance of the introduction of the Anthropocene being implemented in the foreseeable future," believes his colleague Stefan Wansa, the chairman of the Quaternary department of the German Stratigraphic Commission, which is responsible for most recent geological history. "The advocates of the Anthropocene have to put up with the allegation that they are not sufficiently familiar with the rules of stratigraphy," says Wansa.[122]

Geologists like Zalasiewicz, Trombore and Broecker do not take such comments lying down as they are well aware of the rules of their discipline. What's more, Stanley Finney, the world's top stratigrapher, used laudatory words when he wrote at the end of 2013 that the Anthropocene idea

120. Whitney J. Autin und John M. Holbrook, "Is the Anthropocene an issue of stratigraphy or pop culture?" *GSA Today*, 22, vol. 7 (2012): 60–61. http://dx.doi.org/10.1130/G153GW.1.

121. P. L. Gibbard and M. J. C. Walker, "The term 'Anthropocene' in the context of formal geological classification," *Geological Society, London, Special Publications*, first published October 25, 2013; doi 10.1144/SP395.1.

122. Axel Bojanowski and Christian Schwägerl, "Debatte um neues Erdzeitalter: op. cit., see footnote 106.

deserves serious attention.[123] The criticism, however, makes it clear that some geologists would rather cling to a past where a new geological epoch is absolutely not advancing towards us, before our very eyes.

This is why it is so important that the members of the Anthropocene Working Group weigh the facts carefully. If the first vote of the official Anthropocene Working Group returns a positive result in 2016 (of which there is every indication), higher-ranking groups of stratigraphers will then address the question of whether there have been enough geological signals accumulated, differing sufficiently from existing geological strata, that can be measured for a long time to come. But the ultimate decision is not even in the hands of the stratigraphers; it is up to an even higher authority, the Executive Commission of the International Union of Geological Sciences, where there is as much order and discipline as there is in a Swiss wristwatch factory.

It is beginning to dawn on more and more geologists, however, what an opportunity the Anthropocene idea presents. Although the oil and gas industries, not to mention the mining industry, would come to a standstill without geologists, this discipline hardly receives much public attention. This could change abruptly if the Anthropocene idea makes visible how our everyday lives, from our technical gadgets to the consequences of our eating habits, are steeped in geology. In the future, geologists could take on a completely different role in the positive development of the earth, and become "solution providers."

More fundamental criticism comes from humanist academics who are well versed in dealing critically with terminology and questioning the impact on society of particular expressions.[124] Isn't it a dreadful and sexist generalization to speak about an *Age of Man*? After all, humanity consists of billions of individuals who live within thousands of cultures, language groups and traditions. The diversity of emotions, behaviors, thoughts,

123. Stanley C. Finney, The 'Anthropocene' as a ratified unit in the ICS International Chronostratigraphic Chart, op. cit., see footnote 108.

124. I am not referring here so much to published sources as to many valuable discussions that took place during the Anthropocene Project at the Haus der Kulturen der Welt in Berlin, see http://www.hkw.de/en/programm/projekte/2014/anthropozaen/anthropozaen_2013_2014.php.

musical outputs, dreams, actions, and body language has been in competition with biodiversity for a very long time. The difference between an altruist and the perpetrator of genocide or a Facebook user and an indigenous tribal member in Papua New Guinea, or the economic disparity between a starving child in Sudan and an oligarch billionaire in Russia, are enormous. In cultures around the world, there are also radical differences in the perception of time. For some indigenous tribes, like the Aymara, in the Altiplano of the Andes, the past lies in the future and vice versa.[125] Other cultures, like the Pirahã in Amazonia, apparently have no delineations between past, present and future.[126]

There is enormous diversity in the perception and understanding of time as well as the division of units of time.[127, 128] Would the Anthropocene assimilate them all?

It could be argued the word "Anthropocene" declares all humans equally to blame for today's ecological problems. But the responsibility for climate change is unequally distributed. Most carbon dioxide emissions have historically come from Europe and North America. An American produces, on average, sixteen tons of carbon dioxide emissions per year whereas an East Indian produces only two tons. According to estimates by Potsdam University, Oxford University and the World Resources Institute, of the 2,110 billion tons of additional carbon dioxide that went into circulation as a result of human activity between 1800 and 2014, a good half is caused by industrial countries, where only a small portion of humanity lives, while the other half comes from countries which have approximately four fifths of the world's population.[129, 130]

125. Rafael E. Núñez und Eve Sweetser, "With the future behind them: convergent evidence from Aymara language and gesture in the crosslinguistic comparison of spatial construals of time," *Cognitive Science*, vol. 30 (2006): 1–49. http://www.ppls.ed.ac.uk/ppig/documents/NSaymaraproofs.pdf.

126. Daniel Everett, *Don't Sleep, There are Snakes: Life and Language in the Amazonian Jungle*, London: Profile Books, 2009.

127. Chris Lorenz and Berber Bevernage (eds.), *Breaking up Time: Negotiating the Borders between Present, Past and Future*, Göttingen: Vandenhoeck & Ruprecht, 2013.

128. Daniel Rosenberg and Anthony Grafton, *Cartographies of Time—A history of the timeline*, Princeton Architectural Press, 2010.

129. For a graphic display of carbon dioxide emissions, see the project by Oxford University at http://trillionthtonne.org.

130. World Resources Institute, "Navigating the numbers—Greenhouse Gas Data and International Climate Policy," Washington, DC, 2005, http://pdf.wri.org/navigating_numbers.pdf.

Millions of indigenous peoples, like those in Amazonia or Borneo, suffer from the brutal consequences of Western wealth, power and consumerism, including rainforests disappearing before their eyes along with their livelihoods. Such people have to make way for mines and mega plantations so that Smartphones and cheap food can be sold in Europe and the United States. The people of Bangladesh who work in clothing factories, are paid a pittance so that Europeans can buy the latest fashions as cheaply as possible. Wastewater from these textile factories pollutes and dyes the local rivers yellow and pink.[131] These people are "anthropos" too but they are not responsible for these depredations. From this perspective, the word Anthropocene becomes a term of imprisonment for the slum dweller fighting for survival in India, a kind of collective liability for the selfish and wasteful behavior of Americans and Europeans (and more recently, Chinese).

If one defines the Anthropocene as merely the sum total of environmental crimes, then one would have to call it the "Westocene" or "Capitalocene" in reference to the Western lifestyles that have chiefly been the cause of anthropogenic phenomena.[132]

Quite a different perspective opens up if the new geological epoch is more broadly defined, not as the sum of all environmental problems, but instead as a result-oriented project. The Anthropocene could become a kind of forum in which all cultures have equal validity and all people are treated equally, similar to the intent of the United Nations' 1948 Universal Declaration of Human Rights. By citing the "human being" as the agent of the Anthropocene idea, indigenous peoples would be included as modern agents who have equal rights and who play a part in the geology of the future instead of being just victims of anthropogenic change. Indigenous people would no longer be depicted as supporters of a pre-modern, primitive existence, but instead as emancipated contemporaries. This could encourage them to ask for a vote in a global democracy. Seen in this light,

131. Jim Yardley, "Bangladesh pollution told in colors and smells," *New York Times Online*, July 14, 2013, see http://www.nytimes.com/2013/07/15/world/asia/bangladesh-pollution-told-in-colors-and-smells.html?_r=0.

132. The term "Kapitalozän" (Capitalocene) was coined by Prof. Elmar Altvater from the Freie Universität (Free University), Berlin during a discussion at the German Council on Foreign Relations.

the Anthropocene could strengthen opposition to the fact that, at present, a small portion of humanity is altering the climate and biodiversity for its own short-term gain, without any concern for the long-term health of the rest of humanity. In that sense, the Anthropocene would not only be a physical description of the state of things; it would be construed as an ethical demand and guide, the beginning of an awareness-raising process.

Further criticism of the Anthropocene idea comes from environmentalists. Many of them fear that the Anthropocene notion fosters a dangerous technocratic advance, and might be in essence anthropocentric—in other words, that the term suggests the world is being viewed from a purely human point of view, subordinated to human demands. Geologist Stanley Finney asks: "Might the desire to establish the 'Anthropocene' as a formal unit be anthropocentric?"[133] The German biologist Andreas Weber writes: "The Anthropocene position shares with the green economy idea the underlying anthropocentric assumption—that we can (or even must) start from a uniquely human standpoint to come to terms with the problems of sustainability. . . . In Anthropocene thinking, the gap between nature and culture has dissolved, not because humans have come to a different understanding of life and their role in it, but because their technology has swallowed nature."[134]

Sociologist Eileen Crist from the Virginia Polytechnic Institute, in Blacksburg, Virginia, sees the very issue itself at work in the Anthropocene idea, namely a fixation on people. "This name is neither a useful conceptual move nor an empirical no-brainer, but instead a reflection and reinforcement of the anthropocentric actionable worldview that generated 'the Anthropocene'—with all its looming emergencies—in the first place."[135] Another critic, Kathleen Dean Moore, insists that a new geological epoch should be named after whatever comes next, that is, following the environmental catastrophes triggered by humans.[136]

133. S. C. Finney, "The 'Anthropocene,' op. cit. , see footnote 108.

134. Andreas Weber, *Enlivenment, Towards a fundamental shift in the concepts of nature, culture and politics*, Heinrich Böll Foundation, 2013, available here: http://www.boell.de/en/content/enlivenment.

135. Eileen Crist, "On the poverty of our nomenclature," *Environmental Humanities*, vol. 3, (2013): 129-147.

136. Kathleen Dean Moore, "Anthropocene is the wrong word," *Earth Island Journal*, Spring 2013, http://www.earthisland.org/journal/index.php/eij/article/anthropocene/.

Crutzen inadvertently fueled fears of an anthropocentric technocracy in some of his early writings on the topic. In his article "Geology of Mankind," he wrote: "A daunting task lies ahead for scientists and engineers to guide society towards environmentally sustainable management during the era of the Anthropocene. This will require appropriate human behavior at all scales, and may well involve internationally accepted, large-scale geo-engineering projects, for instance to 'optimize' climate."[137]

Scientists and engineers *guiding* society? Is the Anthropocene idea perhaps a vehicle for the rule of undemocratic academics, or even another manifestation of techno-megalomania? Wouldn't an "Age of Man," if taught in schools across the world, strengthen the feeling that the planet is our property to do with—and not do for— whatever we like? An intense debate is necessary to consider such issues. Crutzen should not be judged by his early remarks, nor does he expect his views to be taken as an authoritative canon, but rather as an impetus. He does not even claim to have a precise definition of the Anthropocene, as he once confided to me during a car ride. Under no circumstances does the Nobel laureate envision technocratic large-scale projects, or subordinating the world to humans. Yet, as he said in a long interview with me in 2013: "Ultimately it will be politicians who will have to make some bold decisions to change course. Scientists and engineers can help but their real power lies in making positive innovations possible, not in decision-making."[138]

Crutzen said he would "not apply geo-engineering, at present" and shares the fear that "researching geo-engineering will lead to an attitude that carbon dioxide reductions can be postponed because sulphur injection technology will save us from dangerous climate change." When I confronted him with the criticism leveled by Andreas Weber and others that the Anthropocene was anthropocentric, he was taken completely by surprise because he thinks that the Anthropocene idea is about humans understanding how to live appropriately with the other inhabitants of the Earth. However, he did offer one caveat: "We humans only have our human brains through which to understand the world. So even

137. Paul J. Crutzen, "Geology of mankind," op. cit., see footnote 80.
138. Christian Schwägerl, "Es macht mir Angst, wie verletzlich die Atmosphäre ist" (It scares me how fragile the atmosphere is), interview with Paul Crutzen, *Frankfurter Allgemeine Zeitung*, 20 November 2013.

when you care for the perspectives of other species, you are using a human brain."

In these statements, Crutzen offers hope that it is precisely the anthropogenic perspective, despite its outward similarities to the word "anthropocentric," that could lead to the opposite of anthropocentricism, if humanity understands its central role in the earth's metabolism, and accepts its existential interconnectedness with all other living creatures, as well as with the oceans and their tides, the atmosphere, and the mountains.[139]

Together with John McNeill and Will Steffen, Crutzen has outlined the path to an attitude that makes up the third, imminent phase of the Anthropocene, which he believes will have to start in 2015 at the latest. After the upheavals of the Great Acceleration from 1945 to the present day, people mature into "stewards of the Earth system." "Stewards," as opposed to "Masters," has a thoughtful and considerate ring to it. It suggests that Earth might yet not be irreversibly transformed in a short space of time but, handled with care so it could be handed down to others. This view gives the Anthropocene a meaning not only for scientists but for everyone. It connects global events to everyday occurrences, from the great timescale of geology to the minute by-minute rhythms of daily life.

Of course, the Anthropocene is not an ultimate theory that solves all problems. Quite evidently, the idea must be thoroughly examined from all angles, before school children open their history and geography books to learn that we are living in the geological Age of Human Beings. It is perfectly fine for stratigraphers to take a few years to settle these important questions and not come up with a definitive "yes," just yet. The Anthropocene would nevertheless continue to be a strong metaphor for the fundamental way in which the relationship between humans and time is changing: after all, it is evident that today's perception of time is problematic. We need a new relationship with how we as individuals, groups, nations, and as the whole, use the resources of deep geological history, as well as how we make decisions about the future of our planet and its residents, including all other living beings, present and future.

139. Paul Crutzen and Christian Schwägerl, "Living in the Anthropocene: Toward a New Global Ethos." Essay, New Haven, CT: *Yale Environment 360*, (2010), see http://e360.yale.edu /feature/living_in_the_anthropocene_ toward_a_new_global_ethos_/2363/.

Our lives are bound by these long-term relationships. Even the White House has come to this conclusion. In January 2013, President Barack Obama surprised the world in his second inaugural speech by mentioning an American responsibility "to all posterity" in terms of climate change.[140] This radical assessment may not yet be reflected in policy but nevertheless, one of the most powerful politicians in the world is alert to the timescale with which we are dealing.

The chronological horizon upon which we have a direct influence in our everyday lives reaches at least as far as the next 75,000 years[141, 142]; a timeframe that the scientist Buffon thought was the maximum imaginable age of the Earth. Some changes that humans are causing at present, from atomic waste storage to interventions in evolution, move into the range of *kalpas* or aeons. In addition to the expansion of time that took place in the minds of scientists like Buffon, a massive expansion of time period awareness has taken place, in which our decisions have a direct effect. We can say we have begun to communicate with the deep future—what is our message?

140. http://www.whitehouse.gov/the-press-office/2013/01/21/ inaugural-address-president-barack-obama.

141. Curt Stager, *Deep Future—The next 100,000 years of Life on Earth*, New York: Thomas Dunne Books, 2011.

142. David Archner, *The long thaw—how humans are changing the next 100,000 years of Earth's climate*, Princeton University Press, 2010.

FIVE Apocalypse "No"

I SPENT MY CHILDHOOD on a small "island" between the Soviet Union and the United States of America. I come neither from Alaska nor the Aleutian Islands but from Bavaria, in Germany. When people hear the name Bavaria, they think of *Lederhosen* (leather shorts) and the *Oktoberfest*— but my experience in that part of the world wasn't quite so comfortable. Just twenty kilometers east of my hometown, the "Iron Curtain" ran along the border between West Germany and Czechoslovakia, replete with all its watchtowers, barbed wire fences, minefields and machine-gun nests. The border felt like a black hole that drained away vital energies. On many maps, Czechoslovakia was simply shown as a blank space, without place names, streets or forests, as if it were a void.

And twenty kilometers west of my hometown was Grafenwöhr, at the time the largest American military training area in Europe. At night, the western sky often glowed orange as troops lit up the area with flares that soared into the sky and then drifted slowly down on parachutes. The boom of rapid-fire cannons and windows rattling due to constant gunfire and exploding bombs was as normal as the background music in a café.

My father was an avid hunter and I often went on walks with him into the dark pine forests of our small island between the world's superpowers. One day, we crossed a street and I noticed a manhole cover that looked slightly different from the ones I was accustomed to seeing. I must have been about ten or eleven years old and, out of curiosity, I asked what it was. Had I an inkling of the answer, perhaps I would never have asked.

"Those are shafts for nuclear warheads," said my father. "If the Russians invade, the Americans want to use nuclear bombs as a barrage, first to blow up the roads and then the entire area."

That was my personal "Armageddon moment." The end of the world was suddenly close at hand. For the rest of the walk, I no longer saw trees or deer or the chanterelles, instead, I saw only mushroom clouds from atomic bomb explosions.

From then on, I often imagined what it would actually be like if everything I knew and loved; my family and friends, forests, and idyllic villages were destroyed by an almighty explosion.

During the Cold War, Western and Eastern Bloc armies faced each other with enormous arsenals of tactical nuclear weapons. The fear was that these weapons might be used. If you are young and only know this era from hearsay, you cannot imagine how absurd it was to sit in school and be told to rehearse crawling under our desks, to protect us from the blast of a nuclear attack. Those of you who are old enough to remember, no matter where in the world you are from, may have had your own "Armageddon moment" when the realization dawned on you that the civilization into which you had been born could be obliterated in seconds.

There were plenty of tense incidents in that period that might have led to a nuclear war, such as false alarms at early-warning facilities, or the building of the Berlin Wall by the East German regime, or the Cuban Missile Crisis. On September 26, 1983, when a Soviet early-warning system indicated that an American nuclear missile attack was underway, an experienced Russian lieutenant-colonel named Stanislav Petrov prevented the outbreak of full-scale nuclear war. His system registered a lone, inbound American missile but Petrov knew that only an all-out attack would make sense, so he deduced that it was a false alarm and decided not to respond.[143] Had he decided otherwise, there might have been thousands of nuclear bomb explosions within hours. It might have been the end of civilization and even perhaps the end of three billion years of evolution and life itself.

Between July 1945 and November 1989, there were about 2,000 nuclear tests across the world; these tests have left traces of radioactive isotopes that geologists of the future could use as markers of the early Anthropocene.[144]

---

143. David Hoffmann, "I had a funny feeling in my gut", *Washington Post*, 10 February 1999, see http://www.washingtonpost.com/wp-srv/inatl/longterm/coldwar/shatteo21099b.htm

144. On this subject, see also a timelapse video of every nuclear explosion since 1945 by Japanese artist Isao Hashimoto http://www.youtube.com/watch?v=LLCF7vPanrY.

Characteristically, Paul Crutzen was one of the first scientists to calculate how a nuclear war could lead to a nuclear winter by producing a thick layer of dust in the atmosphere that would block out the sun's light, essential for plant life, and thus killing it off.[145, 146]

But humankind survived the Cold War. Today, a wide freeway goes from my hometown toward Prague, capital of the Czech Republic. The European Union has swallowed up former Warsaw Pact countries from behind the Iron Curtain. In the fight between the superpowers, which was a clash of economic and political philosophies, the will to survive and a thirst for freedom prevailed. Unfortunately, global conflicts are still a real threat. That's why an international public holiday would be a fitting way to mark the fact that humanity was smart enough to avert a nuclear apocalypse. This experience is a good starting point for whatever follows.

What I felt that day in the forest with my father, may be one of the reasons why the Anthropocene so appeals to me. The Anthropocene is an anti-Apocalypse idea, par excellence; an "Apocalypse No" instead of an "Apocalypse Now." After all, it connects the past to the future. From my perspective as a child, my future had shrunk to a virtual speck under the threat of nuclear war, whereas the Anthropocene idea expands almost to infinity. A geological epoch that stretches a mere 250 years since the start of the industrial revolution would hardly be noteworthy. The Anthropocene idea opens the horizon to the next 2,500 years or the next 25,000, if not the next 250,000. For me, the Anthropocene is like a vantage point rediscovered: It creates the prospect of a deep future, of changes for the better. If the Anthropocene consisted only of devastation and environmental destruction, we would need to fight against it and actively prevent it from happening. We would need a global "back-to-the-Holocene" movement—and fast! But if we take the Anthropocene idea seriously, it can help us shape our present behavior in a positive way. Rather than defining humanity as the destroyer of nature, the Anthropocene casts people in an affirmative, long-term role. It is neither about facing an ecological apocalypse, nor harking back to "the good old days." It is about answering

145. Paul Crutzen, Ian E. Galbally, Christoph Brühl, "Atmospheric effects from post-nuclear fires," *Climatic Change*, vol 6, issue 4, (December 1984): 323-364.

146. Paul J. Crutzen und John W. Birks, "The atmosphere after a nuclear war," *Ambio*, vol. 11, (1982): 114–125.

the question of what impact today's population is having on the Earth in the years 2050, or 2500 or 25000. The Anthropocene is not a ticking time bomb, nor is it an end-of-the-world scenario: rather, it is a beginning-of-the-world scenario.

The grave threat of the Cold War strengthened the resolve of millions of people to protect the environment. I have come across people all over the world who feel this way: anti-nuclear-power protestors in Germany, anti-logging activists in Clayquot Sound on Vancouver Island, Canada, biodiversity preservationists in Africa's Congo and rainforest conservationists in Brazil. I always come away from these encounters with positive feelings, due to people's love for the environment and its animals and plants, as well as their concern that the next generation be able to inherit and enjoy the vast richness of life. But a negative attitude that I've often noticed, especially in Europe and North America, is a certain tendency toward apocalyptic views. How many times has it been claimed that "time is running out," or if this or that doesn't happen, it will be too late and the world will be lost forever? A particularly striking example of this is the "Doomsday Clock" that has been maintained since 1947 by the Science and Security Board of the *Bulletin of the Atomic Scientist*.[147] This clock had a legitimate role during the Cold War but, since 1989, it seems as if the scientists involved have been striving to invent new reasons to keep their project going instead of donating the clock to a museum.

Doomsday rhetoric is often used with the best intentions—namely, to rouse politicians and the general public into action. But at the same time, it encourages a dangerous mindset: Doomsday thinking prevents us from imagining a long-term future.

A prime example of modern apocalypticism comes from Professor Stephen Emmott, Head of Computational Science at Microsoft, who published a 200-page pamphlet in 2013 containing many pictures and some text to illustrate the ecological problems we face, in which he arrived at two conclusions: "The problem is us," and "I think we're fucked."[148]

It would be dangerous if a large number of people were to think like Emmott, Most of us human beings would not plant an apple tree today if

147. http://www.thebulletin.org/timeline.
148. Stephen Emmott, *Ten Billion*, op. cit., see footnote 38.

we knew that tomorrow was going to be the end of the world, as Martin Luther put it.[149] Faced with this situation, it seems better to eat apple pie, to enjoy life and consume, one last time, everything that the world has to offer. Perhaps it is no coincidence that this kind of short-term thinking is particularly prevalent in the United States where, more than anywhere else in the world, belief in an imminent apocalypse is actively fostered by Evangelical Churches and movie blockbusters in the style of Roland Emmerich. Nowhere is the New Testament Book of Revelation of St. John quoted more often than in America.[150, 151] In general, economic systems in the very wealthiest nations are geared to short-termism, from shopping on credit to high-speed trading on the world's stock exchanges.

Of course, there is a motive behind the doomsday sermons of those like Stephen Emmott. They sell well because they engender fear and fascination, in equal measure. Something in us loves an Apocalypse—maybe it's the narcissistic feeling of being a member of a chosen generation.

But the Apocalypse gurus seeking attention and making money by frightening people are mistaken. In all probability, the earth will not be destroyed in the foreseeable geological future, at least not in an apocalyptic sense. This means that we humans of today and our descendants will have to live with the long-term consequences of our present actions, good and bad. Even if climate change turns out to be worse than scientists at the IPCC fear, it will not lead to the end of the world or the collapse of civilization. This won't even come to pass if climate change, food crises and cyber wars all occur simultaneously. Without a doubt, there are terrible things in store for humanity if we don't change course but not an apocalypse nor a "we are fucked" doomsday scenario. Those truly interested in preventing negative developments should consider that predicting a worldwide ecological collapse is more likely to persuade people to use up

149. This is an oft-cited quotation, but in fact it appeared for the first time in 1944 in a circular letter from the Hessian Protestant Church, in an attempt to give church members hope.

150. See, for example, the statement by Tea Party politician Michele Bachmann, October 2013: http://www.washingtonpost.com/blogs/on-faith/wp/2013/10/08/michele-bachmann-obama-administration-support-of-syrian-rebels-proves-we're-in-the-end-times/.

151. For a more up-to-date interpretation of the Book of Revelation, see Elaine Pagels, *Revelations. Visions, Prophecy, and Politics in the Book of Revelation*, Viking Penguin, 2012.

the earth's existing resources as quickly as possible rather than husband them wisely over the long-term. What appears to be environmental protectionists' greatest strength, i.e., their dire warnings about the end of the world, provokes a contradictory reaction: We go on a kind of global, closing-down spending spree.

Too much pessimism isn't helpful because doomsayers usually depress people. I became thoroughly despondent in 2009 during an interview with Dennis Meadows, one of the co-authors of the bestseller, *The Limits to Growth*. After our meeting, it took me a long time to shake off the apocalyptic visions he'd evoked.[152]

In my view, the Anthropocene idea pierces through apocalyptic predictions repeatedly stated in countless ecological discussions since *The Limits to Growth* was published and first predicted collapse in the year 2000, then 2020 and 2050.[153] Almost all such analyses predict a future crash because they extend current trends and do not allow for any new phenomenon arising from innovation or social renewal and technological change.

The Anthropocene idea could, however, lead to a "Growth of the Limits"—not in the sense of continual economic growth—but in the sense of social, cultural, spiritual, scientific, and economic renewal. This was the argument behind an editorial in *Nature*, which stated that Crutzen's idea could eventually lead to a new global ethos: "It would encourage a mindset that will be important not only to fully understand the transformation now occurring but to take action to control it."[154]

The German Government's Advisory Council on Global Change sums up this approach in its report: "World in Transition." The Anthropocene not only refers to effects that humanity has had on the earth but also to the cognitive changes undergone by a global civilization that has an increasing awareness of itself as a driving force. And therefore, a "new era of

152.   Christian Schwägerl, Interview with US economist Dennis Meadows: "Copenhagen is about doing as little as possible," *Spiegel Online*, December 9, 2009.

153.   Donella H. Meadows et al., *The Limits to Growth. A report for the Club of Rome's Project on the Predicament of Mankind*. New York: Universe Books, 1972; and see also 'Humanity Is Still on the Way to Destroying Itself', interview with Dennis Meadows in Spiegel Online, http://www.spiegel.de/international/world/limits-to-growth -author-dennis-meadows-says-that-crisis-is-approaching-a-871570.html.

154. Nature Editorial Board, 'The Human Epoch,' *Nature*, vol. 473, (19 May 2011): 254.

responsibility is beginning."[155] Science historian Jürgen Renn describes the Anthropocene as a "process that reflects upon itself."[156]

Defining people as the problem, as Stephen Emmott does, hardly helps. We need positive stories that guide the way ahead.[157] How can an Age of Humans develop if it is driven by a negative attitude toward humans? "Anthropophilia," as Andrew Revkin has called it, is essential for improving the way we treat the planet. I do not have a romantic, idealistic view of humanity, nor a belief in the inherent goodness of people. In Berlin, almost every day, I come across memorials and plaques that remind me of the horrors of Nazism and the Holocaust. In very few cities are there quite so many reminders of the cruelty of which people and political systems are capable. Without doubt, our modern world is fraught with crises, violence and conflicts. What can help us overcome them? We must ask ourselves what are the best conditions whereby the good in us will prevail?

By the beginning of the Great Acceleration in 1945, the Nazi regime had lost the world war. Since then humanity has created many institutions and means for positive change. In 1948 the United Nation's Declaration of Human Rights established a standard to which all nations, whether it be China, the United States, or Syria, would adhere, if they engaged in torture, aggression or repression. The mass-murderers of today can now expect to face the International Court of Justice in The Hague, in the Netherlands. Modern societies are also linked by a network of civil organizations that assist the less fortunate and advocate a fairer distribution of global wealth.

Natural calamities or disasters nowadays generate massive waves of foreign aid. Other positive changes are that scientific research is stronger than ever; revealing the internal secrets of atoms, brains and ecosystems and defeating ignorance. Competence in producing new technologies is increasing and engendering solutions that were unknown only a short time ago, from anti-malaria vaccinations to artificial photosynthesis, from 3D printers to quantum computers. The Internet's World Wide Web and

155. German Advisory Council on Global Change, Flagship Report: "World in Transition. A Social Contract for Sustainability," Berlin, 2011.
156. Speech during the opening of the "Anthropocene Project" at the Haus der Kulturen der Welt in Berlin, January 2013.
157. Jo Confino, Sustainability movement will fail unless it creates a compelling future vision, The Guardian, November 23, 2013, http://www.theguardian.com/sustainable-business/sustainability-movement-fail-future.

social media, which make global connectivity possible, overcome geographical barriers to communication and enable people to participate in social and political processes, no matter where they reside. A teenager living in Africa can take an engineering course over the web at MIT, the Massachusetts Institute of Technology in Cambridge, Massachusetts. And for all its shortcomings, the United Nations is an organization that brings nations and cultures together under one roof, for fair exchanges of views.

Does all this seem naïvely idealistic? At the least, it should sound anthropophilic! If you were to adopt Emmott's apocalyptic vision, you would have to see each newborn baby as an additional burden, instead of congratulating the parents. You'd have to ask who, of the people around you, is one too many for the planet. In Emmott's view, there are too many young people on the earth. This view is flawed. What about young Boyan Slat from the Netherlands, born 1994, who at nineteen designed a ship that can scoop up garbage from the ocean?[158] Or Jack Thomas Andraka from the United States, born in 1997, who at age fifteen developed a test that screens for liver and lung cancers at their very earliest stages?[159] Or Malala Yousafzai, from Pakistan, born in 1997, who was shot and nearly killed by the Taliban because she supported education for girls, who has survived and started a worldwide campaign to that end?[160] Or Anjali Appaduri from India, a delegate at the United Nations Climate Summit, in Dubai, who at twenty-two gave a rousing speech on reducing carbon dioxide emissions?[161] All the same, you don't have to be a child prodigy to get a "Welcome to Earth" greeting card.

A basically positive attitude towards other people is one of the most important elements in the Anthropocene. If people and states assume that others, but not they themselves, are burdensome for the Earth, this will become a further source of hatred and violence. It isn't necessarily a bad thing if the earth becomes more populous. The *anthropization* of the world could lead to a deeper humanization.

"Apocalypse No" means believing in people and a long-term future on earth. We are not intrinsically Nature's enemy, as many environmentalists

158. http://www.boyanslat.com/bio/index.html.
159. http://jackandraka.net/.
160. http://www.malalafund.org/.
161. Anjali Appadurai, "Get it done", www.vimeo.com/33410262.

want us to believe. Instead, we are the medium through which life becomes aware of itself and transforms into something new, in a conscious way, not undirected evolution but by "a process that reflects upon itself."

This reflecting process has not yet matured. What our collective actions lack is long-term rationality or responsibility "to all posterity." Short-term thinking is the currently dominant worldview as we dump our waste to alter the earth of tomorrow, change the climate with our carbon dioxide emissions, and saturate the oceans with incredibly harmful tiny plastic particles.[162] In the Western world, a cult of "now" has sprung up, which believes little in a tomorrow or the day after tomorrow, in terms of apocalyptic doctrine. Life in the present economic system is organized as if there is no tomorrow; in any case, no morning after the end of the next quarter or the current electoral cycle.

This manifests in dreadful paradoxes. We build an increasingly complex infrastructure but at the same time increase our emissions of greenhouse gases thereby increasing the occurrence of extreme weather events. We place more and more value on personal health and longevity but destroy the biological treasure of tropical rainforests, which contain priceless medicinal plants and pharmaceutical active ingredients. We measure our lives and success in terms of money but attribute no economic value to the natural infrastructure that keeps us alive. We value instant global communication in the form of the Internet but at the same time ignore the organic communication networks of ecosystems upon which we are completely dependent.

The situation would be different if we already had a perfect understanding of the climate system, by way of long-term research and we were pumping billions of tons of greenhouse gases or cooling substances into the atmosphere in order to alter the climate, specifically to prevent a new Ice Age. It would be different if we had already invested a good portion of our wealth into recording and understanding the natural systems all around us. We would then have a comprehensive catalog of all the animal and plant species on earth and would know what their reciprocal effects are. Perhaps one could justify eliminating an entire species, in such

162. Christian Luhmann, "Temporal Decision-Making: Insights from Cognitive Neuroscience," *Frontiers in Behavioral Neuroscience*, vol 3, no. 39, (23 October 2009).

a scenario, if it posed great danger to another life-form. At present, this is all done blindly and unconsciously, without thought for the planetary consequences.

In this early phase of the Anthropocene, our dominance over the earth has developed more quickly than our ability to reason and be responsible. Our economic system relies on old thinking. Part of our evolutionary animal nature is our brains' preference for short-term rewards: "A key motivator of our lives, pleasure is central to learning, for we must find things like food, water and sex rewarding in order to survive and pass our genetic material to the next generation," writes the neurobiologist David Linden.[163] When prehistoric hunters and gatherers found a bee hive dripping with honey, or when they hunted down an animal, it was only right to eat as much of the food as quickly as possible, as there was no telling when a rival might appear or where the next food source would be. In the modern world, such genetic programming has disastrous consequences because we are constantly enticed by products high in sugar, protein and fat. The food industry shamelessly exploits the fact that our brains are hardwired for short-term rewards. Agricultural policy leads to negative consequences like obesity and diabetes as well as the clearing of rainforests and savannahs in order to produce animal feed, palm oil and sugar cane.

Such short-term policies are deeply anchored in the operating system of today's economies. A preference for the present economic system leads to shorter intervals between crises.[164] According to this doctrine, a forest that is felled today is inherently worth more than a forest that might still stand 100 years from now. The preference for the present skews all decisions that favor posterity.

One consequence of such attitudes is the way in which corporations are driven to be profitable over shorter profit cycles—three months or less— caused by high-frequency trading in the financial markets, measured in

163. A terrific read: David J. Linden, *The Compass of Pleasure—How our brains make fatty foods, orgasm, exercise, marijuana, generosity, vodka, learning and gambling feel so good*, Viking, 2011.

164. Christian Gollier und Martin Weitzman, "How Should the Distant Future be Discounted When Discount Rates are Uncertain?" Cambridge: Harvard University, informal discussion paper, 7 November 2009 and Christian Gollier et al., "Declining discount rates: economic justifications and implications for long-run policy, Working Papers, LERNA, University of Toulouse, http://ideas.repec.org/s/ler/wpaper.html.

milliseconds, which means computers at multinational banks are programmed to move twelve-figure sums around, of their own accord, in order to maximize profits. "If you see a stock price on your iPhone or computer, it's as if you're looking at a star that has been extinguished for 50,000 years," says a stock broker, explaining the advantage big investors have with high-speed computers.[165] The economic systems of today have replaced, on a grand scale, modes of living that took years or decades to develop, like the sustainable harvesting of Brazil nuts in the rainforest. In their place, palm oil is grown on deforested land, due to profit driven cultivation methods.

Short-term thinking manifests itself in such phenomena as shopaholism and the compulsion for constant electronic distraction. When long lines form outside stores the moment the latest version of a mobile telephone hits the market, we can see this infantile addiction to instant gratification at work. Western mass media is skilled at manipulating the brain's reward center into seeing unnecessary products as essential to survival. Shopaholics behave as if there is no tomorrow.[166] Even in situations where people plan for the future, today's economic system is rooted in a short-term ideology. The colossal sums of money that international pension funds pushed onto capital markets led to the property and financial crises of 2008, with bad loans and "virtual" profits.

Another grim aspect of short-term thinking exists away from the world of plenty. Millions of people have no access to clean drinking water, sufficient nutrition or electricity. Such people are stuck in a poverty trap, from which even the most hard-working and resourceful rarely escape. This forces them into a daily fight for survival, leaving little time to think about the future. A logger is scarcely able to appreciate the long-term value of the tropical rainforest if, while cutting down a valuable hardwood tree, he has no other means of feeding his family.

It is increasingly clear that short-term thinking and denial of interest in the future is not a good path through the twenty-first century. In order to continue our current lifestyle, we have to incur ever greater ecological and financial debts, and take all kinds of bigger risks. Western democra-

165. Frank Schirrmacher, *Ego—Das Spiel des Lebens*, Blessing, 2013.
166. You'll find numerous appaling videos of "Black friday riots" online.

cies have difficulty proposing long-term thinking in their political outlook. Four or five year electoral cycles and the influence of lobbyists with short-term goals have favored decisions that affect and damage the environment. Political and economic systems are not geared to the wellbeing of the earth's future inhabitants. This will change only if a broad coalition of new priorities, new majorities and new political styles prevail, not from fear of an imminent Apocalypse but instead, with a long-term view.

Long-term ideas or aims have been lacking as to how we should manage the massive reorganization of the biosphere that is taking place. Restricted on one hand by neo-liberal short-term thinking and on the other by fear of an ecological Apocalypse, scenarios for a positive, long-term future have fallen by the wayside. Such idealistic scenarios must take into account individual and collective decisions made in everyday life, as well as events taking place in cities and communities or the activities of societies and associations, political parties as well as global forums and committees like the United Nations.

The Anthropocene, as a "process that reflects upon itself," is also about how poverty can be eliminated; how we as people would like to manage the world's climate, and how we can create beautiful and life-sustaining ecological footprints.[167] It's also about how we can increase cultural and biological diversity, connect the infrastructures of civilization and nature and put respect for all life at the center of economic management.

This may sound like an idealistic wish list but we are transforming earth so it seems that wishes on an equally grand scale are required. Aside from that, you can be certain that, as you read these lines, the powers that have caused so many crises during the Anthropocene—powers that are short-sighted, power-hungry and profit-driven—are working through long, much less transcendental wish lists in their strategic planning departments.

Saying "no" to the Apocalypse not only means opposing short-sightedness but also recognizing that the huge crises and upheavals of our times offer equally great opportunities. If up to three billion more people are going to be living in cities by 2050, it means that half of the urban

---

167. See also William McDonough and Michael Braungart, *The Upcycle: Beyond Sustainability—Designing for Abundance*, North Point Press, 2013.

infrastructure has yet to be built. If energy consumption per capita increases, it will go hand in hand with massive energy system investments that could be used to help bring renewable resources to the fore. And if the demand for food rises sharply, there is a chance for alternatives to industrial agriculture to develop. So "Apocalypse No" means visualizing a new kind of future. Currently the "American Dream" still dominates—a dream of unlimited material comfort and happiness, without thought of the consequences. Nowadays, however, the American dream is becoming an Asian, South American and African dream. China has come up with the expression "the Chinese Dream," which resembles the American Dream to its core.[168, 169]

The problem with the American Dream is that it cannot be globalized in a sustainable way. According to ongoing research conducted by the Sustainable Europe Research Institute (SERI), Americans consume 90 kilograms of resources per capita a day, twice as much as Europeans, whose standard of living is already very high. They release 16 tons of carbon dioxide per capita, per year, from energy use alone. They eat more than 200 grams of meat a day. Due to the global demand generated by these patterns of consumption, an increasing number of nightmarish places are springing up: deforested areas in Asia and South America, oxygen-deprived waters in the Gulf of Mexico, bleached coral reefs along tropical coasts, oceans devoid of any fish and homogenous suburban areas proliferating on the outskirts of Western cities, the destruction of Africa's biodiversity, decapitated mountains in Virginia, pock-marked fracking expanses in Texas, tar sand expanses in Alberta, Canada, and the Great Pacific Ocean Garbage Patch.

If everyone followed the old American Dream, with its predilection for big cars, monoculture and enormous quantities of waste, it would result in a massive and long-lasting cultural and ecological impoverishment. What kind of dream is it that becomes a nightmare for everyone trying to live it?

168. Helen Wang, *The Chinese Dream: The Rise of the World's Largest Middle Class and What It Means to You*, Bestseller Press, 2012.

169. In China the definition of the Chinese Dream is disputed. See for example the articles and speeches at http://www.chinadaily.com.cn/china/Chinese-dream.html and Peggy Liu, "China dream: a lifestyle movement with sustainability at its heart," *The Guardian*, June 13, 2012, http://www.theguardian.com/sustainable-business/china-dream-sustainable-living-behaviour-change.

A new dream of the future is needed, or a collection of dreams that can be Chinese, African or European, and can be lived by everyone in his or her own way, increasing cultural and biological diversity instead of eliminating it.

Crises represent times of change, renewal and reorganization. What humanity has already achieved as a community—its agriculture, cities, technologies and science—demonstrates immense artistic and creative power.

Quintessentially, people are learners. So why shouldn't a planet dominated by living people not evolve well? The human race has produced the most astonishing scientists, artists, spiritual leaders and community organizers. Why shouldn't it be possible for today's people to overcome short-term thinking, avarice and economic mismanagement? Resourceful people fill our world with the most amazing machines. Why can't our machines be calibrated to protect earth's life systems rather than exploit them?

In the Anthropocene the future will not become shorter, it will get longer. Medical and sanitary advances are visible manifestations of this process. The average life expectancy of a child born today in the Western world now reaches into the twenty-second century. According to recent demographic studies, half the girls born in industrial countries today will live to be over 100, if present trends continue. Life expectancy is increasing in many regions of the world, at an incredible speed—five or six hours a day. (Although I wouldn't blame you if you don't believe me, I assure you this is not a misprint.)[170] Despite some glaring exceptions, including North American women from poor backgrounds, the overall trend is positive.[171] On a worldwide scale, life expectancy at birth has increased from 63 to 70 years since 1990.[172] Underpinning this increase is progress in medicine

---

170. Jim Oeppen and James W. Vaupel, "Broken Limits of Life Expectancy," *Science*, vol. 296, (2002): 1029-1031, and James W. Vaupel, "Biodemography of human ageing," *Nature*, vol. 464, (25 March 2010): 503.

171. Grace Wyler, "US Women Are Dying Younger Than Their Mothers, and No One Knows Why," *The Atlantic*, (October 7, 2013), http://www.theatlantic.com/health/archive/2013/10/us-women-are-dying-younger-than-their-mothers-and-no-one-knows-why/280259/.

172. http://www.who.int/gho/mortality_burden_disease/life_tables/situation_trends_text/en/index.html.

and education. Biologists and health experts are looking into aging and cell regeneration. In disciplines as diverse as stem cell research or educational psychology, knowledge is growing as to how aging can be decelerated. In some countries over-80-year-olds are the fastest growing segment of the population. Whereas at the end of the nineteenth century, the average life expectancy in Europe was less than 50 years, it is far from absurd to speculate that breakthroughs in knowledge and positive changes in lifestyle could soon make a life span of 120 years possible for many people. This means it would only take six people for a story in the present day to be passed down, person to person, and retold in the year 2500. A girl born today, who reaches the age of 100, could tell the story to her 5-year-old granddaughter, who in turn could tell it to her granddaughter in 2200, and so on.

What this means is that it becomes increasingly difficult to "pass the buck" of climate change or biodiversity degradation or unfair distribution of wealth on to the next generation. The likelihood that these consequences will actually damage the *current* generation is growing with every additional hour of life expectancy.

In the United States, a project designed to assist our comprehension of time is currently in the making. The "Clock of the Long Now" will tick once a year, strike every hundred years, and emit a special tone every thousand years.[173] In the Sierra Diablo mountains in Texas, construction of a full-scale model has already begun. Presenting a longer timescale, the clock is a nice expression of "Apocalypse No." The Long Now clock implies that we aren't the last people on earth, nor a selected generation that will cause its collapse; instead, many generations ahead will look back at us as we have looked back at our ancestors.

"Apocalypse No" means that we are the prehistoric race "of the future." Undeniably, there won't just be a few tools, bones or scripts left behind by our lives, but quadrillions of objects, human-made geological signals and strata. People of the future will be able to reconstruct a very precise image of who we were and what we did for or against them.

With a bit of luck, something special will form part of this legacy— the 12,500 kilometer-long (7,767 miles) strip of green land that runs north

173. http://longnow.org/clock/.

to south through Europe, from the Barents to the Black Sea. The border of the former Iron Curtain, where I began to fear the Apocalypse, is now being transformed into a long and narrow but continuous nature reserve. Because minefields made the area inaccessible to people, rare animals and plants species found a habitat there. This amazing international nature reserve will be maintained as a Green Belt and a contribution to a Green Infrastructure that serves to help all kinds of living creatures spread out across human living spheres.[174] An icon of existential threat is turning into a manifestation of ecological farsightedness, an example of how to say "no" to the Apocalypse.

174. Christian Schwägerl, "A Long Scar from the Iron Curtain, a Green Belt Rises in Germany," *Yale Environment 360*, (April 4, 2011). http://e360.yale.edu/feature/along_scar_from_iron_curtain_a_green_belt_rises_in_germany/2390/, see also http://www.europeangreenbelt.org/.

SIX The Evergreen Revolution

INITIALLY, THE PILOT OF THE SMALL CESSNA aircraft made an effort to steer us around the billowing smoke. He was trying to spare his passengers from having to breathe in trees, butterflies and orchids that had been transformed into carbon monoxide, carbon dioxide, and nitrous oxide. But after a while, he gave up. We just flew straight through the dark clouds of smoke—the remains of a tropical rainforest that had quite literally dissolved into thin air. Our eyes and throats hurt from the aggressive gases.

The Indonesian island of Borneo is one of the hotspots of global forest destruction.[175, 176] In the region east of the provincial capital of Palangkaraya, fire was blazing in every direction as I flew over in the airplane to get a picture of the situation. Not only were the trees ablaze but the forest floor itself, which consists of peat moss and plant remains, was burning, turning biomass into greenhouse gases.[177] The last remaining trees lay like pale skeletons on the ground. East of Palangkaraya, the devastation extended across a good sixty-by-sixty-mile radius, crisscrossed by dead straight drainage channels. Something fundamentally wrong had happened here.

The forests of Borneo are being cleared not just to obtain wood, but primarily to turn the area into agricultural land. Ash from the inciner-

175. L.M. Curran et al., "Lowland forest loss in protected areas of Indonesian Borneo," *Science*, vol. 303, no. 5660, (2004): 1000-1003.

176.   James Morgan, "Forest change mapped by Google Earth," BBC News, November 14, 2013, the inter-active tool http://earthenginepartners.appspot.com/science-2013-global-forest.

177.   Guido van der Werf et al., "CO_2 emissions from forest loss," *Nature Geoscience*, vol. 2, no. 11, (November 2009): 733-808.

ated rainforest trees is used as a kind of start-up capital, fertilizer to grow a product that is found in processed food all over the world: palm oil. It is the cheapest of all edible vegetable fats. When you buy cookies, candy, or convenience foods from the supermarket, you will see palm oil inconspicuously listed as "vegetable oil" on the list of ingredients. Fifty-five million tons of palm oil were produced worldwide in 2013, mostly in Indonesia and Malaysia, with staggering consequences for people and the environment.[178, 179] Although Indonesia placed a moratorium on creating new arable land for palm oil cultivation in 2011, it is not being systematically enforced.[180]

The clearing of forests in the lowlands of Borneo is by far the crudest form of food production that I have ever seen. But I have also witnessed elsewhere the surface of the earth being converted into monocultures for human demand: the clearing of Brazilian woodland to grow soy for European animal feed, the creation of wheat deserts in the USA, or the transformation of biodiverse East German pastureland that is being plowed up to plant fast-growing corn for milk and meat production.[181] Between 2002 and 2012 alone, 2,3 million square kilometers (888,000 square miles) of forest were destroyed, a large portion of that having been assigned to new agricultural zones.[182] Demand for food is the single largest driving force behind deforestation.

The challenge of feeding the growing numbers of people on the planet is the most fundamental collective task in the Anthropocene. The land required to do this is the ultimate source of all modern life, a wafer-thin layer of fertile soil atop Earth's rocky crust. When food becomes scarce, the already thin veneer of civilization threatens to crack: nothing is worse than people fighting over food or wasting away from famine. Many of

178. United States Department of Agriculture, Foreign Agricultural Service, "Oilseeds - World Markets and Trade," September 11, 2014, www.fas.usda.gov.

179. Douglas Sheil et al, "The impacts and opportunities of oil palm in Southeast Asia, Bogor," CIFOR, 2009.

180. D.P. Edwards and W. F. Laurance, "Carbon emissions: Loophole in forest plan for Indonesia," *Nature*, vol. 477, no. 33, (2011).

181. Jan Willem van Gelder et al., "Soy consumption for feed and fuel in the European Union," Profundo Economic Research, (2008).

182. M.C. Hansen et al., "High-Resolution Global Maps of 21st-Century Forest Cover Change," *Science*, vol. 342, no. 6160, (15 November 2013): 850-853.

today's conflicts have implicitly to do with agricultural land or access to water.[183] Should we fail to solve the major issue of global nutrition, a dark Anthropocene epoch will dawn.

The problems ahead are formidable. In addition to the seven billion people who already need sufficient food each day, by 2050, there will be an additional two billion human beings, chiefly in Asia and Africa, who will also want to have enough to eat. The Food and Agricultural Organization of the United Nations (FAO) estimates that if present trends continue, the production of food will have to grow by seventy per cent by 2050 to cover the needs of the global population.

But today, world nutrition is already significantly imbalanced. Although sufficient calories are produced globally each day, roughly 840 million people still go to bed hungry every night, many of them small-scale farmers.[184] Contrast this with industrialized and developing countries where 1.5 billion people suffer from obesity as a result of consuming too much meat, fat and sugar. Land and energy requirements for food production are colossal. But, according to the FAO, of the almost five billion tons of food produced annually, at least a third is not eaten, but thrown away instead—the equivalent of tossing 750 billion dollars a year into the trash.[185]

The job of ensuring that everyone has enough to eat and industrialized populations have a balanced, healthy diet is already a sufficiently major challenge. But, by turning the global landscape into farmland and pastures and forming an agricultural system shaped by economics and technology, humanity has created a further Anthropocene challenge. Farmland and pastures are no longer small islands in an ocean of nature that supplies water and nutrients and absorbs waste. Instead they are the dominant form of land use world-wide.

The huge fields and plantations of corn, wheat, soybeans, rice, cotton and palm oil that extend across the world might seem as if they are "outside" of nature. A farmer or a large corporation decides what is planted,

183. Andreas Rinke and Christian Schwägerl, *11 Drohende Kriege—Künftige Konflikte um Technologien, Rohstoffe, Territorien und Nahrung*, Bertelsmann, 2012.

184. FAO, IFAD and WFP, "The State of Food Insecurity in the World 2013: The multiple dimensions of food security," Rome, *FAO*, 2013.

185. FAO, "Food wastage footprint—impacts on natural resources," Rome, 2013.

which pesticides and fertilizers are used, and everything is then harvested for transport to cities and centers of consumption. But the feeling of being "outside" is an illusion, an historical artifact from the era of colonizers and conquests, when agricultural areas had to be wrested from wilderness, mile by mile, in order for people to survive. As soon as these cultivated fields became the dominant vegetation over large areas, they turned into a new kind of nature. With this "second" kind of nature, there are completely different conditions than for freshly conquered deforested ground. Agricultural systems must not only be sustainable and preserve the "nature" surrounding them, they have to *be* natural and support themselves as ecosystems. This is not simply an end in itself but a prerequisite for securing the future food supply in the long term.

Thus, there is no "environment" any longer that surrounds our civilization. We are living in an "invironment," a new nature that is strongly shaped by human needs and that has no outside.

In this new reality, nature becomes culture and culture becomes nature; technology becomes the environment and the environment is turning into the technosphere; the economy will become ecology and ecology will become economy. In this light, destroying the last remains of real wilderness, like transforming tropical rainforest into a monoculture in the manner that I'd witnessed in Borneo, has become a complete anachronism. Setting alight self-sustaining ecosystems in order to plant industrial food crops is likely to seem quite primitive and barbaric in the future.

Modern agriculture is an impressive testimony to the ingenuity, pioneering spirit and sheer power of humanity. Since the Green Revolution of the 1960s and 1970s, agricultural production has increased faster than the world population. Food prices have fallen noticeably over a long-term perspective. The program of more food per hectare and more hectares of food has been successful in a way that would make a Mesopotamian farmer from the Fertile Crescent, the cradle of agriculture 10,000 years ago, turn green with envy.[186]

The corn yield per hectare in Iowa, for example, increased fivefold between 1936 and 2006; the wheat yield in France increased threefold between 1960 and 2007. In the 1960s, more than half of the world's

186. FAO, "World agriculture—towards 2015/2030," Rome, 2002.

population subsisted on less than 2,200 calories a day. In 2010 this was only true of less than a fifth of all people. China has made the most progress by far.[187]

All this has been a huge success. In 1798 the British economist Thomas Robert Malthus predicted that the population would grow more quickly than the means of agricultural production. He foresaw terrible famines. But generations of agrarian reformers, scientists and farmers have disproved Malthus.

It would be impossible to feed today's population of seven billion people with the manually operated tools used in agriculture in 1798. Either we would not have one square centimeter of unplowed earth left; or the majority of people alive today would simply not exist (probably including you and me).

Malthus, however, underestimated innovation. In the mid-nineteenth century, Justus von Liebig discovered that nitrogen, phosphorous and potassium are critical to plant growth, and that if just one of these elements is insufficient, development may be damaged. His major work, *Die organische Chemie in ihrer Anwendung auf Agricultur und Physiologie* (Organic Chemistry in its Application to Agriculture and Physiology), was the starting signal for a massive increase in yield. At the beginning of the twentieth century, the German chemist Fritz Haber and the businessman Carl Bosch developed a process to convert atmospheric nitrogen and hydrogen into ammonia, a crop fertilizer. Because converting nitrogen into crop fertilizer meant much more food could be grown, the Haber-Bosch process is responsible for feeding billions of people that lived in the twentieth century. This invention has also fundamentally changed earth's nitrogen metabolism. Nowadays, an average of fifteen kilograms of nitrogen per person are spread annually on farmland in the form of fertilizer.[188] And, through agriculture, humanity releases more nitrogen collectively than all other natural terrestrial processes put together.[189]

After World War II, scientific momentum to boost yields gained speed. Norman Borlaug, a farmer's son and agronomist from Iowa, was commis-

187. FAO and OECD, "Agricultural Outlook 2009–2018," Rome, 2009.
188. FAO, "Current world fertilizer trends and outlook to 2011/12," Rome 2008.
189. Peter M. Vitousek, "Human Domination of Earth's Ecosystems," *Science*, vol. 277, no. 5325, (25 July 1997): 494-499.

sioned by the Rockefeller Foundation in 1944 to breed crops with a higher yield. He crossbred dwarf plants with a major wheat variety. The resulting wheat had thick stems with plentiful seed heads that didn't collapse under the weight of the extra grain. From Mexico, the Green Revolution took off: it was a technological offensive that helped developing countries not only significantly increase their yields but their incomes as well. This would have been inconceivable, however, without another farmer's son, Henry Ford, who mass-produced tractors from 1916 on. At the end of World War II, hundreds of thousands of tractors replaced the armored tanks and trucks that had been made in his factories. Today there are approximately thirty million tractors in use across the world, twice as many as forty years ago.[190]

Agriculture has fundamentally changed the surface of the earth within a few centuries. But the phenomenal success story of providing seven billion people with sufficient food is just as much a story about the failure to learn. The present agricultural system is full of flaws and serious problems. It is far away from being a new nature that is biologically rich and able to sustain itself.

Today's food system is very unstable and sensitive to disasters. Unsustainable practices and market speculation result in recurrent price spikes. According to a report by the United Nations Conference on Trade and Development (UNCATD), food prices in the period between 2011 and mid-2013 were approximately eighty per cent higher than in the period between 2003 and 2008. In addition, long-term risks are increasing. UNCTAD warns that the use of fertilizers has increased eightfold over forty years whereas yield has only increased twofold; agricultural productivity no longer increases by two per cent a year as it had done until now, but by only one per cent.[191]

There are plenty of warning signs. In present-day Iraq where the Fertile Crescent once lay, agricultural production has been mismanaged for centuries. The record harvests of the Babylonian era were only possible with intensive irrigation. The water evaporated leaving behind salt. As a result, large expanses of the cradle of agriculture are still desertified. The

190. FAO Stat, 2010.
191. UNCTAD, "Wake up before it is too late: Make agriculture truly sustainable now for food security in a changing climate," 2013.

desert stretches like a negative world heritage site, a salty, depleted region. Iraq, stricken by dictatorship and war, now has to import 1.4 billion dollars' worth of food annually, including wheat, which originated from this world region. Iraq is a prime example of how bad decisions in the past can continue to have ill effects. Had the Babylonians managed their agriculture better, they might have been able to defend themselves against Asian invaders and maintained their position as the global center of science. World history might have taken a different course: perhaps the descendants of the Babylonians would have discovered America.

Today, we are confronted with a crisis of American-style industrial farming. Just as the Mesopotamians ignored the creeping salinization of their soil, we have not resolved the fatal flaws in our own industrial land use. The size of fields has grown over the past few decades. In many regions, they stretch from one end of the horizon to the other, typically without hedgerows and trees, which has led to a sharp decline in biological diversity.[192, 193] Because the same crops are planted often and every last plant is harvested, the soil has no time to retain or regenerate its fertility. This is compensated for with artificial fertilizers but only to a certain point. Erosion, loss of humus and perpetual use of pesticides deplete the soil quality. Industrial farming means that animals we eat mostly live in large outbuildings where they are fed antibiotics, leading in turn to much higher risks of multi-drug resistant pathogens.[194, 195]

Industrial farming has replaced the careful treatment of land and food with large-scale use of oil and gas: Fossil fuels drive tractors that cultivate enormous fields, produce the artificial fertilizers that have replaced natural nutrients and facilitate long-distance transport by truck and airplane to customers. But this dependence on cheap fossil fuels is risky. When oil

192. P.F. Donald et al., "Agricultural intensification and the collapse of Europe's farmland bird populations," *Proceedings of the Royal Society London, B 7*, vol. 268, no. 1462, (January 2001): 25-29.

193. Devra I. Jarvis et al., *Managing Biodiversity in Agricultural Ecosystems*, New York: Columbia University Press, 2010.

194. Thomas Frieden, "Antibiotic Resistance and the Threat to Public Health," Director Centers for Disease Control and Prevention, U.S. Department of Health and Human Services, Testimony Committee on Energy and Commerce: http://www.cdc.gov/washington /testimony/2010/t20100428.htm 28. April 2010.

195. Interagency Task Force on Antimicrobial Resistance, U.S. Government Agencies: *A public health action plan to combat antimicrobial resistance*, Washington, DC, 2009.

prices rose sharply in 2007/2008, food prices followed suit a short time afterwards. The numbers of people suffering from famine worldwide increased immediately and drastically.

Just as intelligent care of the land has been replaced by fossil fuels, the extent to which modern agriculture is far away from playing the role of a functioning "new nature" is shown in the Midwestern and middle Southern states of the USA. In these agricultural centers of the country, fields stretch all the way to the horizon: nature has to obey the symmetry of the plow and the planners' straight lines. This is where vast quantities of corn, wheat, sugar and soy for animal feed are produced in industrial style, generously supported by subsidies from Washington, DC. And thanks to these subsidies, corn is especially ubiquitous in the form of High Fructose Corn Syrup (HFCS), which is so common in many food products today that it could be said that modern Americans are actually made of the stuff to a large degree.[196] Half of the fertilizers that are spread on America's fields today serve only to compensate for the nutrients that have been lost through erosion. "Soil mining" is the term experts give to this over-cultivation of land.[197] A substantial proportion of the fertilizer drains off unused into the water system. While food from the Midwest is sped off via freeway and railway, rainwater, unobstructed by hedgerows and other natural obstacles, carries off soil and surplus fertilizer into the Mississippi and the Gulf of Mexico. In these places, algae and other living creatures thrive so well on the nutrient intake that they deprive the sea of oxygen until it reaches a tipping point and all life suffocates. Since the 1970s, a biological "dead zone" has occurred every year in the Gulf of Mexico: in 2013, it took up an area the size of the state of Connecticut.[198]

The globalization of industrial agriculture comes with terrible risks.[199] Global dead zones already exist in 760 places in the sea and along coastal

196. Michael Pollan, *The Omnivore's Dilemma—a natural history of four meals*, Penguin, 2006.

197. David R. Montgomery, *Dirt—the erosions of civilizations*, Berkeley: University of California Press, 2007.

198. See the NOAA press release: http://coastalscience.noaa.gov/news/coastal-pollution/2013-gulf-of-mexico-dead-zone-size-above-average-but-not-largest/.

199. Karl von Koerber et al., *Globale Ernährungsgewohnheiten und -trends*, Munich /Berlin 2008: http://www.wbgu.de/fileadmin/templates/dateien/veroeffentlichungen /hauptgutachten/jg2008/wbgu_jg2008_ex10.pdf.

areas. They cover a total area of just over 300 square miles, the equivalent in size of New Zealand.[200, 201] By the middle of this century, the planet will have to sustain nine billion people, sixty per cent of whom shall live in cities. By then, it will be decided which path humanity takes on its way into the Anthropocene. Will the last remnants of wilderness be turned into farmland? Will agriculture become the cause of billions of tons of additional greenhouse gases as consumption of meat products continues to rise? To prevent a shortage of food at home, will influential states secure themselves even bigger agricultural areas in Africa using neo-imperial practices?[202]

Agriculture's anthropogenic metamorphosis into "new nature," its transformation into something that could make up a large proportion of the earth's vegetation, cannot work on the basis of the old principles. This is because industrial farming is adversely different from healthy ecosystems in that it:

- is too dependent on an excessive input of energy
- worsens its own climatic requirements through carbon dioxide and methane emissions
- is not resilient to stress and extreme events
- overexploits its most important resource, soil
- is too biologically monotonous to permanently endure.

If biologists discovered a small sub-ecosystem in the middle of the Amazons run by ants that worked in so wasteful and unprofitable a way as does our industrial agrosystem, they would simply shake their heads in disbelief.

---

200. Robert J. Diaz et al., "Spreading Dead Zones and Consequences for Marine Ecosystems," *Science*, vol. 321, (15. August 2008): 926-929.

201. World Resources Institute: http://www.wri.org/our-work/project/eutrophication-and-hypoxia.

202. See Joachim von Braun and Ruth Meinzen-Dick, "Land Grabbing by Foreign Investors in Developing Countries: Risks and Opportunities," IFPRI Policy Brief 13, Washington, D.C., April 2009; Lorenzo Cotula, Sonja Vermeulen, Rebeca Leonard, James Keeley, "Land Grab or Development Opportunity? Agricultural Investment and International Land Deals in Africa," IIED/FAO/IFAD, London/Rome, 2009; Fred Pearce, *The Land Grabbers: The new fight over who owns the Earth*, Beacon Press, 2012 and "Grain, The land grab for food and financial security," Barcelona: GRAIN Briefing, October 2008.

So what would it mean to cultivate land as "invironment," as part of a new anthropogenic nature? It certainly would not mean the return of farmers yoking their plows to oxen, and striding across fields in nineteenth-century garb.

It would be a mistake to idealize or unduly romanticize agricultural production of the past. Life was not only hard then but, thanks to painstaking research, environmental historians have proven that there were also plenty of environmental problems from the overuse of woodland to dangerous erosion in the Mediterranean region, for example.[203] There is no idyllic state to which the landscape could return either. Those "good old days," depicted on product labels in organic food stores and marketed with great determination, showing all those beautiful farmsteads and chubby-cheeked children, were possible primarily because profits from "dirty" industrialization were pumped into the spotless countryside, creating a temporary impression of rural prosperity.

An agroecosystem of the future that plays the role of "new nature" has to fulfill new requirements: it has to meet most of its own energy and nutrient input through self-contained cycles, has to be capable of functioning in demanding climate conditions and only lose so much soil as is it regains. Farming in the future can no longer be based on monocultures, but must be conceived as agroecosystems, providing habitat and food for a number of animals, plants and people. Studies show that productivity and efficiency rise when cultivated land is close to intact biodiverse areas. Coffee plantations near tropical rainforests thrive because pollinating insects fly in from these regions, and the forest provides a good microclimate.[204] New cultivation methods have to be developed, taking into account factors like how favorable it is to plant trees on fields of crops, a practice known as agroforestry.[205, 206]

203. Joachim Radkau, *Nature and Power: A Global History of the Environment*, Cambridge University Press, 2008; Hansjörg Küster, *Geschichte der Landschaft in Mitteleuropa*, C.H. Beck, *1999*.

204. Taylor H. Ricketts et al., "Economic value of tropical forest to coffee production," *Proceedings of the National Academy of Sciences*, vol. 101, no. 34, (2004): 12579-12582.

205. Alain Atangana et al, *Tropical Agroforestry*, Springer, 2013.

206. Gérard Buttoud, Advancing Agroforestry on the Policy Agenda—A guide for decision-makers, Rome: FAO, 2013, http://www.fao.org/docrep/017/i3182e/i3182e00.pdf.

A general change for the better in today's global food system becomes more likely if:

- consumers consciously use their power over production and look for alternatives to industrial diets rich in meat and sugar
- corporations and governments adequately invest in agricultural research and mobilize the knowledge of biology for agriculture
- small-scale family farmers, not industrial firms, are put at the center of agricultural policy
- new methods of food cultivation arise in cities, like rooftop farming or hydroponics.

The United States government spends between 10 and 35 billion dollars annually on agrarian subsidies, and the EU spends nearly 60 billion euros. In addition, subsidies of 500 billion US dollars are spent annually to make fossil fuels cheaper.[207] This money helps to perpetuate bizarre practices. German pig breeders are subsidized to import soybeans from cleared rainforests in Brazil, feeding the soy to their livestock in factory farms that produce vast quantities of surplus food that ends up in the meat being exported to China.

Subsidies in the future should no longer be paid to produce surplus food but to employ farmers as the stewards of a long-term, productive form of agriculture. Every form of state support should be linked to a pledge to maintain soil health, increase biodiversity and produce food in a biodiverse landscape instead of agrarian deserts. The aim should be to preserve cultivated landscapes rather than create industrial landscapes. The EU has already begun this "greening" of subsidies, albeit much too timidly.

Whether the food situation will come to a head at the beginning of the Anthropocene, or whether a change for the better takes place, rests crucially on what people want to eat. A direct link runs from our shopping baskets and plates to the rest of the world. Consumer demand commands

207. Ron Nixon, "Billionaires received U.S. farm subsidies, study finds," *New York Times*, November 7, 2013, http://www.nytimes.com/2013/11/07/us/billionaires-received-us-farm-subsidies-report-finds.html and George Eaton, "Revealed: how we pay our richest landowners millions in subsidies," *New Statesman*, September 19, 2012, http://www.newstatesman.com/blogs/politics/2012/09/revealed-how-we-pay-our-richest-landowners-millions-subsidies.

enormous sums of money, and food producers and traders in particular react very sensitively to the wishes and criticisms of their customers. Whatever we order is a daily choice about the kind of world we want to live in, and it all adds up to a gigantic, collective agricultural production program in which meat and milk and palm oil all play major roles.[208, 209] Because animals need to be fed for a long time before they can produce milk or be processed into meat, and because they lose a lot of energy as body heat, the energy cost for one animal-derived calorie is many times higher than for one plant-derived calorie. Vast areas are cultivated solely to produce animal feed such as wheat, corn and soy for meat production. A large proportion of the deforestation in the Amazonas leads back to the expansion of soybean farming.[210]

But milk and meat consumption does not have to be demonized in the way that some orthodox vegans believe. On the contrary, cattle farming, if it is done right, can even enrich the soil and increase biodiversity.[211, 212] But this means that farmers keep the cattle on the land in limited numbers and that they know a lot about ecological cycles. The amounts of meat produced that way could only be much smaller than what factory farms churn out.

The FAO forecasts that by 2030, one billion additional farm animals will be necessary to meet foreseeable demand. If this additional food is produced by current methods, additional greenhouse gases, slurry, erosion and dead zones will follow. Currently, every kilogram of dairy products produced causes an average of 2.4 kilograms of carbon dioxide in emissions, and every liter of milk is equivalent to burning 0.33 liters of oil.[213]

208. Henning Steinfeld et al., "Livestock's long shadow—environmental issues and options," Rome: FAO, 2006.

209. Swedish National Food Administration, "The National Food Administration's environmentally effective food choices," Stockholm: proposal to the EU Commission, 15 May 2009.

210. For one particular and very local example, see Christian Schwägerl, "Reviving Europe's Biodiversity By Importing Exotic Animals," Yale Environment 360, (January 10, 2013). http://e360.yale.edu/feature/reviving_europes_biodiversity_by_importing_exotic_animals /2608/.

211. D.M. Broom et al., "Sustainable, efficient livestock production with high biodiversity and good welfare for animals," Proceedings of the Royal Society, vol. 280, no. 1771, (November 22, 2013).

212. FAO, "Greenhouse Gas emissions from the dairy sector," Rome, 2010.

213. On this subject, see also Anthony McMichael, "Food, livestock production, energy, climate change, and health," The Lancet, vol. 370 (6 October 2007): 1253-1263.

The average German eats 88 kilograms (194 pounds) of meat products per year compared to an average 100 kilograms (220 pounds) in France and 120 kilograms (265 pounds) in America. The global average has risen from twenty-five kilograms (55 pounds) in the 1960s to almost forty-two kilograms (93 pounds) today.[214]

Eating meat isn't in itself an ecological sin, but excessive consumption on the current scale is. Europeans and Americans could try to achieve the average Asian's consumption of only thirty-one kilograms (68 pounds) of meat a year as a starting point. A sustainable agriculture can only be gradually built if people all over the world consume significantly less meat than they do today, perhaps twice a week, but certainly not on a daily basis, nor as an ingredient of nearly every meal. If meat consumption increases in countries like India or China as people adopt Westernized lifestyles after decades of very low meat intake, consumption in other countries *must* decrease.

A more recent alternative is in vitro meat generated from stem cells in laboratories. This doesn't create entire animals, but simply those parts that humans like to eat. It's a field that certainly requires further research, especially the aspect of whether in vitro meat truly uses fewer resources than animal meat. Another aspect is whether eating meat manufactured in a laboratory would further alienate us from the living world. In any case, animal rights activists seem to welcome this plan.[215, 216]

Meat, of course, can taste delicious. But the less often you eat it, the more you can look forward to it. The money saved from meat consumption could be put into higher-value food products. Farmers who cultivate rather than over-exploit their land, investing in renewable energy and paying attention to high quality, would profit. Contrary to all predictions from agroindustrial interest groups, small-scale farmers in developing countries were able to cultivate their land effectively and in environmentally friendly ways, so long as they were given access to modern knowledge

214. UNEP, Growing greenhouse gas emissions due to meat production, October 2012, www.unep.org/pdf/UNEP-GEAS_OCT_2012.pdf.

215. Alok Jha, "Synthetic meat: how the world's costliest burger made it on to the plate," *The Guardian*, August 5, 2013, http://www.theguardian.com/science/2013/aug/05/synthetic-meat-burger-stem-cells.

216. The animal rights organization PETA has even offered a 1-million-dollar prize for the first synthetic chicken meat: see http://www.peta.org/features/vitro-meat-contest/.

and key markets.[217] To this end, in 2013 the UN organization UNCTAD called for a shift placing small-scale farmers and family businesses at the center of agricultural policies.[218]

Products made by such small-scale farmers certainly cost more in the short-term than cheap industrial foods. But low prices in mega markets are only possible because the true costs are being passed on to the general public and the future. Of course milk, palm oil and coffee can be produced at ridiculously low prices. But the results will be the same the world over: a much higher bill will have to be paid in terms of erosion, eutrophication, and loss of biodiversity and climate change.

Residents of the Anthropocene can change the planet with every meal. Depending on what they eat, they can either expand or shrink agricultural areas and industrial farms. Countryside all over the world can either be deformed into monotonous deserts, or become cultivated landscapes. Poison and slurry sprinklers can either open, or close. Rainforests can either remain standing or burn down. Every single meal is a ballot about these questions.

Dutch researchers have quantified this: "Even small changes in food consumption patterns can have large impacts on the agricultural area required to produce this food. For example, in The Netherlands a hot meal mostly includes some meat, potatoes, rice or pasta, and vegetables. A slight increase of the consumption of meat by only one mouthful (10 g) per capita per day will increase the agricultural area required by 103 square meters per household per year (+3%), whereas the same increase of potato consumption will result in an increase of only two square meters per household per year (+0.05%)."[219]

Examples of genuinely cultivated landscape can be found all over the world. Whether in the Swiss Alps or the Lüneberger Heath, the Polish

217. Ivette Perfecto and John Vandermeer, "The agroecological matrix as alternative to the land-sparing/agriculture intensification model," *Proceedings of the National Academy of Sciences*, vol. 107, no. 13, (2010): 5786-5791.

218. UNCTAD, "Wake up before it is too late: Make agriculture truly sustainable now for food security in a changing climate," 2013.

219. Winnie Gerbens-Leenes, "Consumption patterns and their effects on land required for food," *Ecological Economics*, vol. 42, (2002): 185-199, and Winnie Gerbens-Leenes et al., "A method to determine land requirements relating to food consumption patterns," *Agriculture, Ecosystems and Environment*, vol. 90, no. 1, (2002): 47-58.

marshes or French vineyards, the impressive results of human creation can be admired. But, like the rice terraces of Bali or the remarkably diverse cultivation of mountainous land in the Andes, most of these were created by conditions of hard necessity and are not sufficiently productive for today's requirements. Nonetheless, these cultivated landscapes are an expression of human capability to transform landscapes in a positive fashion. This is highly relevant for the anthropogenic nature of the future.

The first green revolution was about quantity. Now, a second, truly green, "evergreen" revolution is needed, focusing on quality and health.[220, 221] The need for research is immense if agriculture has to act as earth's new nature. A field is a kind of analog Internet, a complex system made up of microorganisms, catalytic surfaces, worms, networks of fungi, aqueous solutions, and insects, all of which compete, cooperate, and work together. Although arable networks represent the oldest anthropogenic ecosystems, they are still largely unexplored by scientists. They are as unknown to us as remote regions of the cosmos.[222] The same holds true for the microbes in our guts. This "microbiome" is pivotal to how healthfully we live, but is largely unknown. Intensified research into agriculture and nutrition is urgently needed: How do soil microbes interact with the plants we want to eat? How can microbes help overweight people find their way back to healthy eating habits? How can the transition from monoculture to biodiverse agriculture succeed? What needs to be done so that small-scale farmers can cultivate their land economically?

One of the first agricultural academies in Europe, today's University of Hohenheim in Stuttgart, was founded following the eruption of the Indonesian volcano Mount Tambora in 1815, which led to global cooling and severe famines. These days, there is a much bigger geological phenomenon than a volcanic eruption—the Anthropocene, and this demands sustained and growing investments in agricultural research and development.

220. Uma Lele et al., "Transforming Agricultural Research for Development," *The Global Forum for Agricultural Research* (GFAR), 2010.

221. John Beddington et al., "Achieving food security in the face of climate change: Final report from the Commission on Sustainable Agriculture and Climate Change," *CGIAR Research Program on Climate Change, Agriculture and Food Security* (CCAFS), Copenhagen, Denmark, 2012. Available online at: www.ccafs.cgiar.org/commission.

222. The current status of research is very well summarized in: Dania H. Wall et al, *Soil Ecology and Ecosystem Services*, Oxford University Press, 2013.

Many traditional environmentalists today see bio-organic farming as the only solution to the problems that we are facing. There is no doubt that organic farming provides us with vital stimulus because it relies on closed cycles of materials without the addition of fertilizers and pesticides, and places a high value on biodiversity. To be able to successfully cultivate land organically with minimal use of technology, a great deal of knowledge is required. But if intelligently practiced, organic farming can produce even higher yields than industrial farming, as shown by examples from many parts of the world, including India, Ghana and Europe.[223] But, as yet there is no conclusive answer as to whether seven billion people can be sustained using today's organic farming methods. Simply relying on the current knowledge of today's organic farmers, seems to me somewhat risky, almost as risky as continuing the practices of industrial farming.

I believe it is time to rethink some of our well-worn negative stereotypes and artificial divisions. Organic farmers are not primitive Romantics, and genetic engineering, if done by the right people under the right circumstances, is not necessarily bad. Fundamentalist opponents to genetic engineering are clinging to a notion of untouched nature that has long since ceased to exist in the growing of arable crops. It is not even an unnatural process to convert the genetic makeup of a crop; plant viruses do this all the time. It is more a question of what intentions lay behind carrying out these changes and who is making money from them.

Today's genetically modified plants are not particularly forward-looking. Instead of assisting real biological solutions, they facilitate an even harsher use of herbicides. The oligopoly of a few large international corporations that produce them is not compatible with a diverse world of millions of small-scale farmers. These farmers should be sowing as many different crops as possible to minimize the risk of pest infestations, rather than being dependent on a small number of crops, varieties and suppliers.

But *must* genetic engineering remain forever synonymous with industrial farming? Or can scientists develop this technology to serve more careful genetic and bioengineering purposes that will help the next gen-

223. FAO, "World agriculture—towards 2015/2030," op. cit. See footnote 184.

eration of ecologists create new varieties of cultivated landscapes? Could some kind of "high-tech" organic farming emerge from a fusion between organic and high-tech agriculture, in other words, from two spheres that have previously been antagonistic? Could "high-tech" organic farming use nature's genetic treasure trove effectively but for completely sustainable ends?

We have seen that modern industrial farming fails to function as "new nature" because today's methods are too primitive. The biocides that randomly destroy weeds and insects render visible just how little the agricultural industry knows about productive interaction and biological control of pests.

There is huge potential for "high-tech" organic farming. Such an approach could exploit a significantly higher number of the 250,000 known plant species for human nutrition, energy supply and engineering materials than the current handful of dominant species. Scientists could take advantage of natural genetic variations that protect plants and animals from climate stress and transfer them into cultivated species GPS-controlled tractors could plow around curvaceous landscapes enhanced by hedges, ponds and small groves instead of in straight lines. Purpose-bred leguminous trees could absorb noxious substances and supply the soil with nutrients via root bacteria.

But "high-tech" organic farming is hardly possible so long as agroindustrial firms like Monsanto dominate agriculture. This corporation excels at aggressive business practices. It tries, for example, to profit from the fact that the wind naturally distributes its patented gene constructs. The ancestral right of famers to save seed from the harvest to sow the following year is a practice that Monsanto would like to see prohibited worldwide. This business behavior, which is based on subjugation, has brought genetic engineering into disrepute for good reasons.

However, Monsanto has only been able to succeed because resources for state-funded scientists at agricultural universities and research institutes are too scarce. What serves gene monopolists above all is the paucity of state research. If universities were in the lead, they would be in a position to place patent rights within foundations and collective trusts. This would be the beginning of an authentic open source biology, a biotechnological democracy that would replace the kingdoms of genetic engineering that are in the hands of shadowy individual firms. A general public that refuses to contemplate genetic engineering per se is likely to wake up in a world dominated by genetic monopolists. Only when that public recaptures and develops the field of genetic engineering for itself will there be a chance for "high-tech" organic farming to succeed.

With improved seed types and methods of cultivation developed by universities available as open source data, it might be possible for small-scale farmers and family businesses to return to the center of the global agricultural system. To a large extent, the industrial farming system is an extension of Wall Street. It is optimized not for the land, but for those handing out ever larger credits to farmers. A constant expansion of machinery, land mass and sales, is necessary to repay interest in order to take out new loans. This capital intensity is unhealthy as it steers farmers' attention away from being concerned about the land to worrying about the banks. Reducing farmers' dependency on credit is an important goal. This is why it is important that farmers can directly market their produce for prices that reflect the hard work necessary to sustain the land. And this is also why all attempts that more people start producing food are so important. The spread of urban gardening in many big cities worldwide is therefore most promising.

If four billion people are going to live in cities in the future, these cities cannot solely depend on distant fields and pastures. Vegetable patches on high-rise buildings, beehives on roofs, communal gardens in restricted traffic zones, the combined breeding of fish and vegetables in "hydroponic container farms"—the list of projects that are already up and working is long. Mayors across the globe are committed to reviving their cities through urban farming.[224] In the USA, Seattle is leading the way by building an "edible park" in which passersby will be able to help themselves to free fruit; and in economically-depressed Detroit, destitute city districts are being turned into urban farms.[225]

224. For concrete examples in Berlin, see Christian Schwägerl, "In Berlin, Bringing Bees Back to the Heart of the City," *Yale Environment 360*, September 6, 2011: http://e360.yale.edu /feature/urban_beekeeping_berlin_brings_bees_back_to_the_city/2439/ as well as http:// prinzessinnengarten.net/about/ and http://www.ecf-center.de/en/ecf-containerfarm/.

225. See http://www.beaconfoodforest.org/; http://inhabitat.com/detroits-urban-agri-culture-movement-could-help-green-the-city/; http://inhabitat.com/hundreds-of-vacant -detroit-lots-to-become-worlds-largest-urban-farm.

What a contrast such projects are to the industrial wastelands caused by palm oil cultivation that I witnessed in Indonesia. In new urban farms, devoted, motivated people work together to help produce their own food. These projects also aid people with low incomes while making cities more beautiful and alleviating environmental problems. By contrast, the palm oil plantations profit primarily a few owners of large corporations, create monotonous landscapes and are a huge factor in contributing to global warming.

The mood of many people whom I met in Indonesia's burned-down forests was despondent. Before their very eyes, their environment had been completely and rapidly transformed, driven by short-term investment interests and the demand around the globe for cheap food. These people had been displaced from their homes without even having to leave home. "We went fishing every day and had a good life here," said a father of four children who was now short of the money he needed to pay for their education. "We used to spot orangutans and hornbills," enthused the children. The workers who cleared and burned down the forest at least got something to eat for their actions but they have also remained poor. They represent hundreds of millions of people who have been bypassed by the promises of the modern world. They cannot afford to educate their children properly, often become seriously ill at a young age, and live from hand to mouth.

Ironically, even the people for whose needs the rainforest was burned down, often become ill. Palm oil fat as a hidden ingredient in many food products makes millions of people overweight, risking their health and diminishing their quality of life.

The agricultural system of today is a tremendous success because the majority of people have enough to eat. But equally, by destroying its very foundations, it is tremendously insane. If agriculture is to function like nature, humanity has to radically change its course. Thousands of small projects show ways to do this. What is lacking, however, is global momentum.

Close your eyes for a moment and imagine the corn deserts of Iowa replenished with life, with hedges, and patches of forest full of birds between the fields. Imagine South America's and Asia's millions of barren hectares currently depleted by agriculture restored to rich forests and agro-forestry farms that can feed these countries sustainably. How the former "granaries of the world" in the Middle East, now over-salinized, can flourish again.[226] Picture new cultivated landscapes spreading in Europe, replacing animal factories.

All of these changes are possible. They start with the kind of food we put on our plates, day by day.

226. In 2010 before the American troops began to withdraw from Iraq, there were 112,000 soldiers still stationed there—but only thirty-eight American agricultural advisors.

SEVEN

The Invironment

L

OCAL CONSERVATIONISTS HAD DRIVEN ME through the narrow streets of Nagoya, a city with 2 million inhabitants. We had come to a sudden halt in front of a small shrine and old trees that formed something like an entrance gate. The scenery felt enchanted, like in one of the animated worlds of Japanese film director Hayao Miyazaki. Gnarled oaks lined the path into the area. Kaki fruits, guavas and quinces hung from trees. A silver dragon plant peeked out from underneath a bamboo. School children had prepared rice saplings to be planted in the following season. In the district of Hirabari, ten hectares of land forms a *satoyama*—a Japanese concept that describes a rich mixture of countryside and agriculture where people obtain food in harmony with nature.

I had come to this Japanese harbor city for a United Nations conference on biodiversity, to report on negotiations over new aims in global nature conservation. The conference was attended by delegates from 190 countries. But, while environmental ministers and representatives from all over the world were wrangling over billion-dollar sums and vast nature reserves, a highly symbolic fight was going on between Nagoya residents just a few kilometers away.

When we arrived at the heart of the *satoyama* area, we heard the buzz of a chainsaw. The grove was the scene of a showdown, and very little harmony was in the air. Construction workers pushed their way past a small group of protesting conservationists, and along a path at the edge of the woodland. Two men in sunglasses and black suits, whose slick appearances would not have looked out of place in a Japanese thriller, closed ranks behind the construction workers. The chainsaws revved up and the first tree was felled.

A short time thereafter, conservationists from all over Japan gathered in the UN conference grounds to demonstrate for the preservation of the Hirabari grove. They were festooned with ferns, branches and flowers, making them look a bit like extras from the film *Avatar*. "Give nature room!" they chanted as security staff rushed to the scene.

The contrast to what was going on inside could not have been greater. In the conference rooms, negotiations focused on the frantic, yet absolutely vital, attempt to conserve as much "wild" nature as possible. Over the course of tough, all-night disputes, the delegates finally came to an agreement that would put under protection, by the year 2020, seventeen per cent of the earth's landmass as well as ten per cent of its coastal and ocean regions. If this target is reached, it will signify a huge success. Effectively, it would mean that protected areas of land would increase from nearly 16 million square kilometers (6.2 million square miles) in 2011 to 25 million square kilometers (9.6 million square miles) in 2020, and marine conservation areas would increase from the current 8 to over 36 million square kilometers (1.9 and 13.9 million square miles, respectively).[227]

The Japanese demonstrators outside, on the other hand, were concerned about something very different: nature *outside* of protected areas. Even if the ambitious aims of the Nagoya conference are attained, the share of nature that will *not* be within designated parks and reserves, and thus not protected in any formal way, will make up eighty-three per cent of the earth's land surface and ninety per cent of its ocean regions. This type of nature might be a piece of land in a city's center that looks just like a real jungle or a farm—or even a mixture of both—until you realize that it is surrounded by skyscrapers and highways. It is precisely this type of area that will be an important part of the new nature of tomorrow, the nature of the Anthropocene.

Over hundreds of years, generations of Western scholars have struggled to draw clear lines between nature and culture, environment and civilization, ecology and economy. In the past, this clearly corresponded to a deep-seated human need to stand out from the rest of the world, to declare ourselves different from animals and plants, which were regarded

227. See World Database on Protected Areas, www.wdpa.org and www.protectedplanet. net.

as "lowly" and "primitive." When we talk about nature reserves nowadays, we are operating from a different need: we put up signs, draw lines on a map, decide what people may or may not do there, defining the area as "nature" in order to save land from human intervention. But this concept of untouched nature is becoming more and more unrealistic. So the question becomes: what *is* nature?

It turns out that the idea of nature itself undergoes constant change and evolution. Even long before the beginning of the modern environmental movement, there were fierce disputes on how to define the term "nature." In the seventeenth century, the English philosopher and scientist Robert Boyle listed dozens of various connotations of the word and questioned whether the term "nature" was even meaningful.[228] In the eighteenth century, focus was shifted to the organization of life around us. The era of explorers and scientists like Maria Sibylla Merian, Carl von Linné and Alexander von Humboldt had begun, eventually culminating in Darwin's theory of evolution. In the nineteenth and twentieth centuries, ideas of predetermined and unchangeable nature were politically and ideologically abused on a major scale to justify such crimes as slavery, kin liability and genocide.[229] "That little word nature is one of terror's favorites," wrote the German philosopher Günther Anders.[230]

Fortunately, we have lived beyond that era, although philosophers, sociologists and biologists still struggle today to define the borders between nature and culture, both in ideology and in real life.[231]

Mindsets from the past lie atop one another like archaeological layers, and continue to exist in the modern world in oddly disparate ways. The animistic worldview of primeval tribes came from an assumption of spiritual unity between living creatures and the earth, a notion perpetuated today by many supporters of ecological and spiritual groups. Various religious doctrines upheld a view of earth as an act of creation directed by powers on high—a vision to which creationists still stubbornly cling.

228. Robert Boyle, *A free enquiry into the vulgarly received notion of nature*, 1686, a Cambridge Texts in the Histor of Philosophy reprint. Edward B. Davis and Michael Hunter (eds), Cambridge University Press, 1996.

229. Frank Uekoetter, *The Green and the Brown—a history of conservation in Nazi Germany*, Cambridge University Press, 2006.

230. Günther Anders, *Die Antiquiertheit des Menschen*, Munich: C.H. Beck, 1987.

231. Philippe Descola, *Beyond Nature and Culture*, University of Chicago Press, 2013.

From the beginning of the Industrial Revolution on, a mechanistic view of the world became popular, extolling the engineer as global designer and regarding nature as mere raw material. This school of thought has shaped our modern economic system. And lastly, in environmental debates over the past forty years, humans have featured largely as disrupters and destroyers, inspiring social initiatives, environmental laws and the Green Economy movement. Writers like Clive Hamilton and Elizabeth Kolbert tend to follow the narrative of humans as the great hooligans of the world.[232]

But to what does nature in the twenty-first century refer—what is nature in the Anthropocene epoch? Does it still refer to untouched rainforests? Rainbow-hued coral reefs? Deserted savannahs? Historical research has revealed that time and again, the notion of "untouched wilderness" was often an artifact. As Emma Marris's outstanding book *Rambunctious Garden* describes, North America was already densely populated before the Europeans invaded.[233] Then millions of Native Americans died from diseases introduced by the first waves of conquerors. The new settlers infiltrated deserted landscapes and imagined themselves to be in "untouched nature." Later on, townsfolk arrived who began to idealize the lonesome wilderness.

In modern-day Amazonia, however, these ideas of untouched nature are being questioned: beneath cleared, supposedly primeval forests, traces of previous civilizations are often found.[234] "Nature didn't get her all-natural identity branding until the Industrial Revolution broke out," writes Bruce Sterling, "then poets and philosophers were allowed to live in dense, well-supplied cities, where they could recast nature from some intellectual distance."[235] With great foresight in the 1990s, Daniel Botkin, a biologist at the University of California, warned: "We have clouded our perception of nature with false images; and as long as we continue to do

232. Libby Robins et al., *The Future of Nature*, Yale University Press, 2013.

233. Emma Marris, *Rambunctious Garden—Saving Nature in a Post-Wild World*, Bloomsbury, USA, 2011.

234. Simon Romero, "Once Hidden by Forest, Carvings in Land Attest to Amazon's Lost World", *New York Times*, January 14, 2012, http://www.nytimes.com/2012/01/15/world/americas/land-carvings-attest-to-amazons-lost-world.html?pagewanted=all&_r=0.

235. Koert van Mensvoort and Hendrik-Jan Grievink, *Next Nature—Nature changes along with us*, Actar, New York 2011.

that, we will cloud our perception of ourselves, cripple our ability to manage natural resources, and choose the wrong approaches to dealing with global environmental concerns."[236]

In the Anthropocene there is no longer an "inside" and an "outside," no alien, antagonistic nature with which rational humans are faced. The environment becomes the "invironment." Instead of untouched nature, there is only touched nature. In the nature of the future, humans will encounter themselves and the former results of their previous actions.

The crucial change in our image of nature in the Anthropocene has been formulated by Paul Crutzen, Will Steffen and John McNeill in the following way: "Humans are not an outside force perturbing an otherwise natural system but rather an integral and interacting part of the Earth System itself."[237] Rainforests and coral reefs will forever be very important elements of nature even when taking this view into account but their fate is now tightly bound to human decisions. Rainforests will no longer exist just because they have always existed, but because increasing numbers of people want them to exist. Coral reefs will only survive if humans decide to stop putting so much corrosive carbon dioxide from exhaust emissions and chimneys into the atmosphere, which ends up in the water. If landscapes shaped by humans dominate earth in the Anthropocene— if artificially-bred animals populate the world, if synthetic substances invade the last corners of the planet and raw materials from the ocean floor find their way into supermarkets—we can no longer speak of nature and culture as two separate spheres. Protected areas will still make a lot of sense—in fact, reserves will be our ecological central banks of the future but they cannot be the only refuges of nature.

A new kind of nature is being created, one that is shaped by humanity. It consists of the sum of all the changes caused by humans on earth. One small part of this "neo-nature" was found by geographer Erle Ellis while trekking on Squirrel Island off the coast of Maine. Ellis was relaxing with friends by the sea, enjoying the sandy beach and the roar of the surf, when

236. Daniel Botkin, *Discordant Harmonies, A New Ecology for the Twenty-First Century*, Oxford University Press, 1990.
237. Will Steffen, Paul J. Crutzen und John R. McNeill, "The Anthropocene: Are Humans Now Overwhelming the Great Forces of Nature?", *Ambio*, vol. 36, no. 8, (December 2007): 614-621.

he suddenly spotted a glittering object at his feet that looked like a relic from a mysterious, undersea world. It was a stone about the size of a tennis ball, strangely deformed, as if water had been eroding it for millions of years. Ellis was excited and put his find in his pocket. Only a hundred meters further along, however, his dreams of having discovered a natural wonder were dashed when he came across a small rubbish dump where the islanders had deposited and set afire all kinds of garbage. In his pocket was a melted piece of civilization.[238] This anecdote certainly does not justify burning rubbish, but it shows how the technosphere and the biosphere interact today to form neo-natural phenomena. Another example of this is Glass Beach near Fort Bragg, California. Until the 1960s, this coastal location was used to dump old glass. The constant flow of water over time created a spectacularly colorful beach, which is now preserved as part of MacKerricher State Park. As waves grind down the glass, people are even talking about replenishing it with further waste glass.

For many nature lovers (including me), it is often difficult to relate to neo-nature. But then consider how some of the species that we love so much, like pandas and tigers, are being preserved only through intensive human interference. On our planet, which is now pervaded by human thoughts and actions, wild animals like lions, tigers and snakes have been transformed into domestic creatures of a kind, just with heightened personal space. They join the ranks of the human-loving dog, the distanced cat, the synanthropic white stork that lives on rooftops, and the neurotic city rat. Wherever animals live "in the wild," they increasingly do so because people have decided that they can.

Ultimately, anthropogenic ecosystems extend all the way to highly controlled zones, such as hospitals, where people create their very own evolutionary conditions for bacteria; or industrial zones and waste incinerator plants, where pigeons, rodents, raccoons and new varieties of plants have all found niches. In the future, nature lovers who face the challenges of the Anthropocene will be looking into the ecological potential of business parks, over-cultivated fields, strips of land adjoining highways, balconies on high-rises and treatment ponds—not to mention airports where, on the

238. Lecture by Erle Ellis for the opening of the Anthropocene project at the Haus der Kulturen der Welt in January, 2013 and personal correspondance. See also videos of this and related talks on the HKW Anthropocene channel on Youtube.com.

east coast of the United States, snowy owls have chosen to nest because the terrain resembles their own habitat in the Arctic tundra.[239]

When I was bird-watching recently in the United States, my sites included Montlake Fill in Seattle, a waste dump that has turned into a bird paradise (as described in Constance Sidles's wonderful books), and a former sewage reservoir happily accepted by numerous sandhill cranes near Madison, Wisconsin. At the Post Office ponds adjacent to Chicago's O'Hare International Airport, under thundering planes, I spotted a pair of killdeer. Human intrusion goes even deeper: By the year 2050, even rare orchids in the remotest rainforests will largely consist of carbon dioxide atoms that once made a trip through an exhaust pipe or chimney.

None of this means that the last isolated wilderness spots on earth should be allowed to disappear or be left unprotected. Quite the contrary: the *importance* of whatever is left of the savannahs, coral reefs, rainforests, deep-sea zones, mangroves, high plateaus and sea grass landscapes will increase in proportion to their scarceness.[240] Such areas will soon be regarded as infinitely precious to us as are priceless artworks by Leonardo da Vinci or Picasso. Their economic, ecological and spiritual value will grow substantially. If all goes well, and the promises of the UN summit in Nagoya are kept, the network of protected nature reserves will have substantially grown by 2020.

But due to the sheer global expansion of cities, industrial areas, military training grounds, landfills and infrastructure, things are radically changing. It is a process that is developing in two different directions. Firstly, what we used to refer to as nature is becoming a green infrastructure, as well as the green security system of human civilization. On a planet that is so thoroughly urbanized, Amazonia has come to resemble something like earth's Central Park. From this perspective, rainforests are carbon dioxide repositories; polar regions are air conditioners, glaciers are freshwater tanks, and mangroves and coral reefs are infrastructure for coastal protection. In the other direction, what we used to call culture is becoming

239. Emma Fitzsimmons, "Snowy Owls to Be Trapped Instead of Shot at New York Area Airports," *New York Times*, December 10, 2013: http://www.nytimes.com/2013/12/10 /nyregion/snowy-owls-to-be-trapped-instead-of-shot-at-new-york-area-airports.html.

240. One good example of this is mangroves. See: Norm Duke et al., "A world without mangroves?" *Science*, vol. 317, (6 July 2007): 41.

an integral part of the biosphere. As humans infiltrate nature, nature also infiltrates human systems.

To make this point at the opening of the Anthropocene project at the Haus der Kulturen der Welt in January 2013, the Austrian writer Raoul Schrott turned our usual mindset upside down. The Anthropocene could not just be described as a process in which people appear on the geological timescale, artifice dominates nature, and nature becomes a product of people, he said. Precisely the opposite was also true: the Anthropocene is an epoch in which geology manifests itself in humanity, nature dominates artifice, and for the first time, people appear to be a true product of nature. It seems as if the Anthropocene is a kind of picture puzzle, like the eighteenth-century image of a peninsula by Bohemian illustrator Wenzel Hollar which, when looked at in a certain way, could also be interpreted as a human head.[241]

According to this logic, not only arable land, but also tens of thousands of residential areas and cities like Nagoya, New York, Shanghai, or Berlin are ecosystems that will have to biologically function like rainforests or moors in the future. Seen from this perspective, our machines are inhabitants of earth that somehow have to become part of its metabolism. Our infrastructure, especially marine traffic, is also part of the neo-natural mechanisms by which plants, bacteria, insects and mammals propagate.

Nature and culture, living creatures and technical objects, everything that comes about and everything that is invented, form new hybrids and amalgamations.[242] This is one of the exciting new phenomena of the Anthropocene. One example of this kind of change in the animal world is reflected by the Australian lyre bird that normally imitates other birds in its courtship songs. Since it has had to share its habitat with humans, it has added the sounds of cell phones, camera shutters, and sadly, even buzzing chainsaws to its repertoire.[243] On YouTube there are videos of more eccentric examples of neo-nature: elks playing on trampolines, a

241. See University of Toronto, The Wenceslaus Hollar Digital Collection, http://bit. ly/1AQUfUU.

242. Philippe Descola, *The Ecology of Others*, Chicago: Prickly Paradigm Press, 2013.

243. This is captured in an astonishing BBC video with David Attenborough: http:// www.youtube.com/watch?v=VjEoKdfos4Y.

crow who keeps sliding down a roof on a plastic lid, an eagle who steals a park ranger's camera and shoots a selfie.[244]

The world is moving further and further away from the conservationists' ideal of uninhabited spaces in which animals and plants live as they did some hundreds or thousands of years ago. This is because:

- nature is emerging in new places, mostly in cities
- due to global mixing and climate change, there are no more fixed "natural" ranges for animal and plant species
- nature conservation relies increasingly on intervention and technology.

These trends change premises, principles and priorities. A conflict like the one over the Hirabari grove in Nagoya is suddenly no longer a purely urban issue; it turns into a question of what nature will be like in the future.

Until recently, cities were regarded as the polar opposite of nature: as guzzling, dirty abodes of Moloch. Starting out from this idea, pioneers of the ecology movement, such as the deep ecologist Paul Shephard, developed the misanthropic notion that the whole of humanity should be ghettoized in cities, giving the rest of the planet over to a nature devoid of humans. In the 1970s, Shephard called for a complete demographic shift, whereby the entire human race would colonize coastal regions and change their eating habits to bacterial products and algae. The only permissible way for people to travel into the inner wilderness of the continents would be on foot; this space "could be freed for ecological and evolutionary systems on a scale essential to their own requirements and to human synergetic culture."[245] Even if this position is extreme, it reflects the global longing among ecologically minded people for an undisturbed and uninhabited nature.

244. http://www.huffingtonpost.com/2013/11/05/elk-on-trampoline_n_4215094.html?ncid=edlinkusaolpo00000009; http://www.deathandtaxesmag.com/211135/brilliant-crow-makes-a-sled-out-of-mayonnaise-lid-has-more-fun-than-youve-had-all-winter/; http://grist.org/list/an-eagle-stole-a-video-camera-and-made-this-cinematic-masterpiece/?utm_campaign=socialflow&utm_source=facebook&utm_medium=update.

245. Paul Shepard, *The Tender Carnivore and the Sacred Game*, New York: Scribners 1973. See also *Defending the Earth*, a dialogue between Murray Bookchin and Dave Foreman, Montreal: Black Rose Books, 1991; and Bill Devall, *Deep Ecology*. Salt Lake City, UT: Gibbs Smith, 1985.

On the surface, it could be argued that Shephard was right. United Nations researchers assume that almost the entire growth in the global population by 2050 will occur in cities, meaning that sixty to seventy per cent of all people will, by then, be concentrated in urban areas (which unfortunately will be among the primary victims of anthropogenic sea level rise). In coastal regions in particular, megacities and megaregions are starting to form, with up to hundreds of millions of inhabitants.[246, 247]

The present 3.5 billion city residents all over the world are at the forefront of a new kind of urban humanity. Big cities never stop exerting a pull on people. Every day, hundreds of thousands of rural poor individuals stream into these urban centers. Most of them land in slums, driven by hopes of making it rich. Cities like Jakarta in Indonesia form a geological reality—a sea of houses and huts, a landscape of people, stone, metal and wood that never stays still but constantly expands, creeping into rice paddies, forests and up hills. It is an inorganic creature that feeds people and is fed by them.

But urbanization does not always mean that more space is being freed up for wilderness in the classic sense. Cities are all joined together by more than 30 million kilometers (18.6 million miles) of streets, 600,000 kilometers (373,000 miles) of extended rivers and canals and 1 million kilometers (621,000 miles) of railway track.[248] And urbanites do not live on algae. Trillions of transport routes penetrate deep into the areas where palm oil, meat, wood and much more besides are grown. For a single jar of Nutella, hazelnuts from Turkey, palm oil from Malaysia, cocoa from Nigeria and sugar from Brazil have to be brought to one location.[249] Cities can concentrate people, factories, material wealth and innovative power into the smallest of spaces. In these close living conditions lies great ecological potential because energy and surface area can be used efficiently. But how effective that really is depends on the urban way of life.

246. UN Habitat, State of the World's Cities 2010/2011, Nairobi, 2010.

247. Karen C. Seto et al., "Global forecasts of urban expansion to 2030 . . .," op. cit., see footnote 36.

248. World Bank, World Development Indicators, 2005. Data from 2002;. J. Kubec and J. Podzimek, Wasserwegs, Hanau: Verlag Werner Dausien 1996; International Union of Railways, annual report 2008.

249. http://www.theatlantic.com/business/archive/2013/12/map-all-the-countries-that-contribute-to-a-single-jar-of-nutella/282252/.

Efficiency is certainly not the main characteristic of the urban sprawl that characterizes city development in the United States. In the mid-1980s, the Boston-Atlanta Metropolitan Axis (BAMA) was still science fiction when William Gibson thought it up.[250] But today, the BAMA is a 1,000-mile-long reality as revealed by pictures taken from the International Space Station.[251] Fuelled by cheap petroleum products and cheap property loans, people in American suburbs live in oversized houses surrounding oversized cities in which they travel around mostly in oversized cars. The costs in energy and resources to maintain this lifestyle are enormous, yet it is imitated all over the world. In Jakarta, there were 3 million cars on the road ten years ago, now there are 10 million. All around the Indonesian capital, gated communities with names like "Heliconia" are springing up like mushrooms; the only feature that is *not* reminiscent of American models in southern California or Florida is the vegetation. This is happening all across the globe.

The question remains as to whether the rest of the world is following the wrong models or whether there is enough imagination to think up alternatives. If the rest of the world melted away—the oceans, the landmass, stone and earth's core—leaving just the infrastructure built by humans including undersea cables, an extremely delicate geological sculpture would remain. This sculpture is growing at present by 6 million metric tons of concrete a day or 22 billion metric tons a year.[252] With what kind of materials could this sculpture grow in the future, and according to which principles? And how might it look if it was to function in the long term like a new kind of nature? This is a question that is only secondarily aimed at city planners and architects but is primarily directed to the collective of urban residents who decide—by riding a bike instead of taking the car, for example—how their habitats develop. Local urban policies could prove even more important for the future of our planet than major

250. William Gibson, *Neuromancer*, Ace, 1986.

251. The ISS video shot at night is always a pleasure to watch: between 3:50 to 4:10 minutes, the BAMA can be seen.

252. The European Cement Association: http://www.cembureau.eu/about-cement/key-facts-figures. Concrete consists of one part cement, 0.6 parts water and 6.7 parts stone.

UN conferences on climate, biodiversity and other environmental issues. Cities are more agile, they have to compete for investment and educated residents; new trends, fashions and lifestyles quickly spread from town to town.

The conflict surrounding the Hirabari grove in Nagoya, about whether a small, magical green space can remain in a densely populated city, seems anachronistic when put in this perspective. Areas like the grove should be off limits to developers and be valued by city planners at least as highly as industrial areas; because such spots contain the seeds by which cities of the future could transform into organic nature. In places like the Hirabari grove, city dwellers can see how organic systems draw life energy from the sun, how they build efficient structures such as trees from biological materials, how the most diverse life forms coexist and how waste products can be fully recycled. The organic city of the future should be more similar to the Hirabari grove instead of devouring it.

In organic cities, the focus will be on drawing energy from local and renewable sources for everything, including for transport. Private cars that run on gas will become as outmoded as horse carriages seem today. Public transport, bicycle highways and a commercial rental system for electric cars will be so well-connected that everyone is able to get to their destination with less stress and fewer traffic jams than at present. Architects will build high-rise buildings whose facades, balconies and roofs double up as farms; or green spaces that optimize their microclimate and offer habitat to countless types of species.

There are already many models for these kinds of buildings: self-cooling houses in North African medinas; futuristic façades with zebra stripes that use air-convection cooling; vertical gardens on buildings in Paris; the gardens in Hamburg that produce algae in summer for use as fuel in winter. These types of houses can be optimized for biodiversity: they can be fitted with nesting boxes for owls and falcons or special shafts for bats. Green bridges can link houses to each other, thus helping to create living roof-landscapes.[253] The anthropogenic city is surrounded by biological systems from mangroves to pasturelands that support and protect it, hold

253. My thanks to artists Myriel Milicevic and Alex Toland for the inspiration.

back floods, absorb carbon dioxide and store water; or woods that cool it and provide recreation.

Cities that adopt these kinds of strategies will also experience positive social changes: their high quality of life will attract well-trained crafts-people and experts, their meeting points will be filled with new activities, and shared spaces will bring people from all social strata together.[254] The more organically cities function, the sooner conflicts that are caused by inequality, "nature deficit disorder," and depressing living conditions will disappear.[255]

Just how well this can work is something I often experience in my home town of Berlin. It is not uncommon here to see swifts swoop through the streets—the high buildings are like the cliffs and crags where they bred and hatched their eggs before the arrival of humans. Common mergan-ser ducks hibernate here at special locations where warm wastewater from industrial areas keeps the rivers from freezing even in winter. The rare eagle owl has found a habitat in a shopping district; the once endangered peregrine falcon hatches its eggs in the tower of the city hall, using the light beam from the television tower to hunt prey at night. Rare or endan-gered bird species such as the bittern, red-backed shrike and European stonechat have settled on various airport grounds. For many of these bird species, the city offers a better biotope then the intensively farmed ara-ble and industrial land in the surrounding countryside. There are nature reserves on former railway grounds, urban fallow land containing rare plant species and man-made hills composed of war rubble. All of these vital areas are very popular with Berlin's residents. They offer spaces for recreation to all social and ethnic groups, cool the city in the summer and provide wonderful adventure playgrounds for children.

The Anthropocene imperative that cities have to function like nature was encapsulated by Marina Alberti, Professor for Urban Design and Planning at the University of Washington, when she demanded that cit-

---

254. For some wonderful examples, see also Mohsen Mostafavi and Gareth Doherty, *Ecological Urbanism*, Lars Müller, 2010.

255. Richard Louv, *The Nature Principle—Human Restoration and the End of Nature-Deficit Disorder*, Algonquin Books, 2011.

ies have to learn to "think like planets."[256, 257] At the moment, cities live *off* the planet, but the movement to change this is already gathering momentum.

But cities are only part of the Anthropocene's neo-nature: areas that have been shaped by human activity, been intensively modified and populated in equal measure, are growing in size worldwide even outside city boundaries.[258] Anthropogenic neo-nature can be as exciting as the "untouched nature" that Alexander von Humboldt explored 200 years ago. The difference is that it is brimming with human intentions, a social network par excellence. It can be the result of more or less good intentions, wishes, dreams and ambitions; or the result of mistakes, coincidences and mishaps.

One such mishap is the Salton Sea, an enormous but shallow salt lake in southern California that was created by an engineering blunder in the early twentieth century. Attempts to divert water from the Colorado River for agricultural purposes got completely out of hand, resulting in the river's flow temporarily changing direction until an enormous hollow in the middle of a desert area was flooded. Since then, the Salton Sea has become one of the most important breeding, migration and hibernation areas in southwest America. Green-winged teals, Canadian geese, American avocets, black-necked stilts, snow geese pintails and eared grebes all spend the winter here.[259] There is also a buildup of environmental problems caused by the constant flow of fertilizers and pesticides from surrounding agricultural areas but this wetland area in the middle of the desert would never have been possible without human activity.

Something similar happened near the Romanian capital of Bucharest during the Communist era, when hydraulic engineers attempted to create an amusement park with a harbor on the River Dâmbovita. A massive

256. Marina Alberti, "Cities as Hybrid Ecosystems," http://www.thenatureofcities. com/2013/09/27/building-cities-that-think-like-planets/ based on Paul Hirsch and Bryan Norton, "Thinking like a Planet," in: Allen Thompson and Jeremy Bendik-Keymer (eds.): *Ethical Adaptation to Climate Change: Human Virtues of the Future,* MIT Press, 2012.

257. Marina Alberti, *Advances in Urban Ecology—Integrating Human and Ecological Processes in Urban Ecosystems,* Springer, 2008.

258. Emma Marris, *Rambunctious Garden—Saving Nature in a Post-Wild World,* Bloomsbury, 2011.

259. http://www.desertusa.com/salton/oct_salton.html.

ring of concrete walls was built for it but, because the engineers had mis-calculated the river's water level, the project was never realized. Instead, neo-nature engulfed the enclosed area and a diverse terrain sprung up, reminiscent of a river delta region. Lake Vacaresti, as the area was named for the surrounding neighborhood, furnishes habitat for many animals and plants, and the residents of Bucharest like to stroll within the concrete enclosure to enjoy the view.[260]

Even human waste can create new habitats: sunken oil rigs and ships lying on the seabed can give organisms the chance to attach themselves. Over the years, many of these "artificial reefs" with high biodiversity have been created. In some places, even old railway carriages and army tanks have been deliberately sunk for this purpose.[261] Not all of these projects have been successful, however, as is shown by the Osborne Reef off the coast of Florida. It was formed in 1972 when 700,000 discarded car tires were sunk in the well-meaning hope that fish and other sea creatures would use them as habitat. But the project turned out to be a disaster and the US military has been busy salvaging the tires since 2007.[262]

In Berlin, neo-nature is budding from a very particular kind of human waste: the feces of the city's former inhabitants. In the 1960s, excretions from millions of people were spread all around the village of Hobrechts-felde, on the northern outskirts of the city. For a while, this waste dis-posal system functioned very well; on some of these expanses of human manure it was even possible to grow vegetables and breed fish. But then, too many pharmaceuticals, synthetic detergents and industrial wastes ended up in the sewage farms, so industrial waste treatment plants came into fashion. The area was cut off from the constant water flow that had shaped it. What remained was a massively over-fertilized landscape with a somewhat bizarre appearance. Those unaware of the history of this area stop and wonder at its ruined concrete walls, its mysterious parapets and its bright green vegetation that somehow fails to look natural. A mix of robust grazing animals is currently being introduced here, creating one

260. http://sustainablecitiescollective.com/big-city/184236/strange-case-bucharest-s-lake-v-c-re-ti. A flight over the area: http://www.youtube.com/watch?v=NRz5PvPXBKs.
261.   For example, see: http://www.theatlantic.com/infocus/2011/04/artificial-reefs-around-the-world/100042/ und http://myfwc.com/conservation/saltwater/artificial-reefs/.
262. "Tires meant to foster sea life choke it instead," *New York Times*, February 18, 2007.

of the largest wood pasture landscapes in Europe: white English park cattle, Scottish highland and Uckermark cattle, Konik and fjord horses roam the 800-hectare landscape. With every mouthful they eat and every step they take, they are having an impact on the vegetation. A sparse forest that will function like a savannah is planned. The long-term aim is to make the landscape scenery marshier, more open and greater in biodiversity. While commonplace flora like feathery reed grasses will decline in the process, rarer species of fauna such as stag beetles and whinchats will increase. But planners of this state-funded initiative have no preconceived notions of how the project should develop. As the project manager explained during my visit, this venture is operating on the principle of trial and error.

Similarly eager experimentation can be found happening at two former military training areas near Berlin, Jüterborg-West and Döberitzer Heide. Over the past two centuries, these areas have been used in turn by the Prussian army, the German Empire, the Nazi regime and the Soviet Union. Since the fall of the Berlin Wall and the withdrawal of Russian troops, they have lain fallow. But the full force unleashed by tanks, troops and gunfire achieved an astonishing feat, for beneath the modern-day Brandenburg landscape, a different, older landscape has been revealed, an open sandy vista more typical of the last Ice Age. Its expanses of sand dunes shaped by the wind yet so far from the sea are reminiscent of a miniature Sahara. A bit further on, the scenery is dotted with military legacies such as bunkers, rubble, and junk metal; while sedges, silver grass and birch trees create the impression of a savannah. In Jüterborg-West, soldiers unearthed a landscape that is both primeval and futuristic, a mixture of desert, heath, grassland and wooded areas. This habitat attracts many animal and plant species that have become rare in other areas, from the dune jumping spider (*Yllenus arenarius*) to nightjars.

To prevent the open areas from becoming overgrown, the Environment Foundation has introduced Mongolian Przewalski horses and European bison. Both of these animals have a special history. Mongolian horses went extinct in the wild soon after their discovery in the early twentieth century due to a frenzy to show them off in European zoos. The species only survived in those zoos. Similarly, European bison were hunted to extinction in the wild in 1927. Both species have been propagated in zoos, and today, they are often re-introduced into the wild for grassland conservation projects. In the former military training areas, horses and bison eat the growing bushes, keeping the sandy areas open.

In Jüterborg-West, animal species that have become rare in farmed landscapes take advantage of man-made structures: the hoopoe bird, with its distinctive orange, white and black crown, has been given artificial pipes as breeding grounds. Conservationists have adjusted many of the old bunkers to accommodate bats. Scattered metal plates in the countryside provide nice, warm, sunny spots for smooth snakes, which are otherwise rare in Germany.

No longer serving the military, these training grounds are now part of research studies into how diverse neo-nature can be. The American military is doing similar experiments in Florida, trying to optimize its military training grounds for endangered animal and plant species. The training grounds near Berlin have become training grounds of a different kind. In the middle of a temperate climate zone, a range of heat-loving animals and plants, able to cope with the higher temperatures that will be ushered in with climate change, has found a home. These species that live at Jüterborg-West and Döberitzer Heide have already adapted to future climate changes and from here, they could colonize larger areas of Central Europe.

Animals and plants have always been in flux and the distribution areas of species have always changed dynamically. But the Anthropocene marks a new phase of accelerated change. If the climate does change as the IPCC predicts, then habitats on the entire earth will change massively. Almost all living things are adjusted to a particular temperature range. Generalists, such as humans, who can adapt to conditions from the subzero temperatures of Greenland to the extreme heat of the Sahara, are more the exception than the rule. Many cold-loving fish populations have begun to shift their territories from temperate latitudes toward the poles, which scientists explain as the result of warming seas.[263] On land too, scientists have observed the way in which dispersal sites for different species are being

263. William Cheung et al., "Large-scale redistribution of maximum fisheries catch potential in the global ocean under climate change," *Global Change Biology*, vol. 16, issue 1, (January 2010): 24–35; Stephen D. Simpson et al., "Continental Shelf-Wide Response of a Fish Assemblage to Rapid Warming of the Sea," *Current Biology*, vol. 21, issue 18, (September 15, 2011): 1565–1570; Keith Brander, "Impacts of climate change on fisheries," *Journal of Marine Systems*, vol. 79, (2010): 389–402.

transformed by climate change.[264] Heat-loving species, like some butter-flies, have extended their habitats northwards, while those specialized in inhabiting the highest and coldest parts of mountain ranges face existen-tial problems as they cannot move any further up.

Biology textbooks still show "natural ranges" of habitats for species but with climate change and ubiquitous means of transport, this concept is pretty much obsolete. As significant as climate change are the millions of shifts directly resulting from human activity. Ever since the Silk Road between Asia and Europe and the Columbian Exchange between Europe and the American continent came into existence, people have transported animals, plants, fungi, bacteria and viruses beyond their former "natural" territories. Corn and potatoes, parrots and guinea pigs are outstanding examples of deliberate transfer. I once took three women from Peru, who had lived in Berlin and stayed within the city boundaries for many years, to the countryside surrounding the city. When they saw maize planted on the fields, they started to weep and feel homesick. We had jumped across 500 years of colonial history together, as the women had not been aware that maize had been brought from their homeland to Europe long ago.

In some cases, brutal colonial histories lurk behind the transfer of cer-tain animals or plants such as the rubber tree, which was brought from Amazonia to the Royal Botanical Gardens at Kew in London, where it ulti-mately helped enrich the British Empire with rubber plantations in Asia, while rubber farmers in the Amazon went out of business. Biologically, unintentional transfers are just as influential—mussels that travel the world in the ballast water of ships; bacteria that accompany people in airplanes; or plants that are imported to botanical gardens and their seeds escape.

The consequences of transference can vary enormously. Populations of the red-crowned parrot are endangered in their "natural" native country of Mexico because habitats are being destroyed and many animals are caught for the pet shop trade. But in California, Texas and Florida, animals that have been freed or have escaped from their owners have formed new pop-

264. Camille Parmesan and Gary Yohe, "A globally coherent fingerprint of climate change impacts across natural systems," *Nature*, vol. 421, (January 2, 2003): 37–42 and Gian-Reto Walter, "Community and ecosystem responses to recent climate change," *Philosophical Transactions of the Royal Society B*, vol. 365, no. 1549, (July 12, 2010): 2019–2024.

ulations that collectively add up to thousands. Particularly in California, parrots have become quite common in urban areas without any negative consequences for other species.[265] The picture is quite different on islands where non-native cats, rats and snakes introduced by humans, can cause veritable massacres. One example is the brown tree snake that arrived on the Pacific Island of Guam after World War II and has been decimating the local bird population ever since. Local trees also suffer because their seeds are no longer spread by the birds. For this reason, conservationists on Guam have been trying to suppress reproduction in brown tree snakes, just as other invasive species are commonly being discouraged elsewhere in the world.

Defining the line between native species and invasive species that are either tolerated or fought is enormously difficult. Significantly, we humans have settled all over the planet and transport ourselves by ship or train, car or airplane by the millions across long distances. But when other organisms take these same routes or are driven from their former habitats and look for new homes, they are immediately classified as "invasive" and culled.

I do not want to trivialize problems that arise through invasive species, but I fear that in the Anthropocene, classical definitions of which species are native and which are invasive will no longer work. Human beings have been mixing up the biosphere as no other creatures have before them; this triggers anxieties that all creatures could be present everywhere at some point, and that all biotopes will look the same, amounting to a "Homogenocene," as John Curnutt has called it.[266] However, I consider it more likely that, in the long run, this mixing of the biosphere will lead to the exact opposite: a gigantic, planetary recombination of species, biological communities, and habitats resulting in a new, complex diversity. I am anticipating a "Heterocene"—with positive as well as negative aspects.

The traditional way to ensure preservation of an animal or plant species was to put up a fence around its habitat and make sure that people stayed as far away as possible. In the Anthropocene, people are everywhere, habitats shift and the distribution of animals and plants is globally mixed. Nature

265. This story is thanks to Ursula Heise, author of *Sense of Place and Sense of Planet*, Oxford University Press, 2008.

266. John L. Curnutt, "A Guide to the Homogenocene," *Ecology*, vol. 81, no.6, (June 2000): 1756-1757.

conservation has become nature design. At the same time, the spectrum of conservationist activities has enormously expanded. Breeding programs at zoos and botanical gardens are the last resort for many species even at this stage. Animals like the Przewalski horse, the Kihansi spray toad, or the Arabian Oryx are only present in the wild because they were protected in the past.[267] Plant seeds are being preserved in enormous collections such as the Svalbard Global Seed Vault in Norway. Microorganisms, DNA, animal eggs and sperm are being held in "frozen zoos" like the one at the San Diego Zoo Conservation Research in the United States, where they are stored until they might be reintroduced into the environment or used for human purposes in the distant future. An independent branch of biology known as restoration ecology, is concerned with reviving and restoring damaged or impoverished habitats to their former states.[268]

Genetic engineers are already investigating whether long-extinct animals like the mammoth can be reanimated using cloning technology, and technology on the whole is playing an increasingly greater role. Rare animals are now being monitored by a dense network of cameras, sensors and tracking devices in order to optimize conservation measures.[269] With growing frequency, drones are being used in conservation to uncover illegal logging practices or to detect the presence of rare species.[270] Many decisions on ecological issues are being influenced by the net of optical sensors in outer space that generate real-time images of our planet, its vegetation, chemicals, ocean currents and fault lines with increasingly better resolution.[271] People, nature and technology are merging into one.

When a new United Nations conference on biodiversity takes place in 2020, the most pressing question will be whether the twenty ambitious goals for global conservation that were negotiated in Nagoya have been

267. Lee Boyd, *Przewalski's Horse: The History and Biology,* State University of New York Press, 1994.

268. See Jelte van Andel and James Aronson, *Restoration Ecology : The New Frontier,* John Wiley & Sons, 2012 and the website of the Society for Ecological Restoration http://www.ser.org/.

269. Etienne Benson, *Wired Wilderness: Technologies of Tracking and the Making of Modern Wildlife,* Johns Hopkins University Press, 2010.

270. http://conservationdrones.org/.

271. http://newswatch.nationalgeographic.com/2013/11/25/hi-def-space-selfies-coming-to-your-web-browser-soon/.

reached.[272] It would be a huge success if by then, nations have managed to bring seventeen per cent of the landmass and ten per cent of the sea surface area under some form of effective protection. This would protect animals and plants for their own sake. It would also be good for humans, because in nature reserves, we are protecting ourselves from the consequences of our mistakes. Through reserves we can come to understand how living systems work and make use of these insights for our cities, for pharmaceutical and technological advances.

But even if the Nagoya targets are reached, eighty-three per cent of Earth's landmass and ninety per cent of its oceans will not be formally protected. These areas—the largest parts of our planet Earth—will morph from biomes to anthromes, like the Hirabari grove did long ago. The fate of humans and millions of other species will be linked to how well the new nature *outside* of conservation areas functions. This neo-nature will become an expression of our identity.

272. Aichi Targets the Convention on Biological Diversity: http://www.cbd.int/sp /targets/.

EIGHT Technature

I N THIS NEW EPOCH OF HUMANS, it's not human beings that are new: we *Homo sapiens* have been around as a species for about 200,000 years. Our characteristics, our ways of thinking, feeling and acting, are ancient. Loving and hating, sharing and craving, cooperating and fighting, are primeval phenomena that are deeply anchored in our biology. What really is new in the beginning of the Anthropocene are the tools and machines that humans have been filling the world with for several centuries: first by mining metal ores, heating them, purifying them and finally shaping or building them into complicated gadgets—cars, computers, telephones, factories and now, robots that can construct robots. Technology has mushroomed across the planet and has fundamentally changed the flow of material and information on earth.

Within just 250 years, we have added the technosphere to the geosphere and biosphere and it infiltrates the life of nearly every person on the planet with the possible exception of the last uncontacted tribes in Papua, New Guinea, and Amazonia.

Smartphones and cars are two of the most radical examples of how people in the Anthropocene do not operate alone but are accompanied everywhere by technology developed by and for them. Some people feel incomplete without their smartphones; others feel their phones vibrating even when they aren't (I admit this has happened to me). In a US survey, a surprisingly high number of the people interviewed declared that they would rather spend a night in prison than hand over their smartphones (I intend *not* to get to this stage). People seem to form symbiotic relationships more with this machine than with any other device. When it comes to cars, we merge with them as if in some kind of personal union. Without thinking, we tend to say, "I'm parked over there" to refer to our stationary vehicles; some people feel physical discomfort when a stranger touches their car, and some drivers are more concerned about their paintwork than the lives of pedestrians and cyclists.

Machines have been ubiquitous for some time. Most people begin their day with a machine, earn their livings, and spend their free time with machines. It's not always easy to tell whether machines are the extensions of humans or the other way round. Airplanes and drones are as much a part of airspace as eagles and geese used to be; ships are more common at sea than whales and tuna fish. Worldwide, one billion cars crisscross landscapes like metal organisms. Not a single wild creature of such size exists anywhere in the world in such high numbers. Up in the sky, earth is surrounded by a swarm of military and civilian satellites.

Machines began as replacements for draft animals; they then became independent, forming their very own evolutionary lineages—first in mobility, then in agriculture, communication, war, research, and finally, in building new machines. Since the twentieth century, technological evolution has eclipsed the Cambrian explosion of some 500 million years ago, during which period the wealth of animal phyla proliferated. If geologists in the future search for traces from our times, they will probably find more technofossils than biological ones. As geologist Jan Zalasiewicz and his colleagues point out: "Current evolution of the technosphere, of which the technofossils are the preserved remnant, is hence now orders of magnitude faster than biological evolution."[273]

If you walk down the long shelves of an electronics store and envision the speed at which all those products are replaced, you might feel as if you are in a kind of hyper-dynamic and hyper-diverse technological ecosystem that is in constant change, much like a forest or a coral reef, just not nearly as beautiful. In general, ten minutes after walking into an electronics store, I feel exhausted and sad, whereas ten minutes in a forest makes me feel renewed and happy.

Within a mere hundred years, technology has progressed from one-

273. Jan Zalasiewicz, Mark Williams, Colin N Waters, Anthony D Barnosky and Peter Haff.," The technofossil record of humans," *The Anthropocene Review*, January 7, 2014, http://anr.sagepub.com/content/early/2014/01/06/2053019613514953.full.pdf.

engine small aircraft to jumbo jets that carry 850 passengers; from steam locomotives to 300-mile-an-hour high-speed trains; from the telegraph to the Internet; from simple weather measurements done by hand to a ring of environmental observation satellites; from the magnetic compass to the Global Positioning System; from punch cards to supercomputers; from handheld telescopes to the Hubble Space Telescope; from microscopes to ATLAS, the big-bang simulator at the CERN laboratory; and from simple bombs and guns to nuclear missiles, satellite-driven precision weapons, and drones.

These days, biological *Homo sapiens* live in such tight symbiosis with their own technical gadgets that they hardly register their presence as machines any more. Most new technologies are largely invisible to us, hidden on anonymous server "farms," or in deserted factories full of robots, or secreted on hard drives and in algorithms. The majority of data traffic on the Internet no longer runs between people but between bots.[274] New digital ecosystems are in the early stages of development; but few of us are truly familiar with them or understand them.[275]

In just a few decades, machines have colonized not only earth's surface but also our minds. Machines are much more than just prosthetic extensions to relieve us of physical labor. They are a means of thought, perception, and negotiation. This is the "new perception of the world" that technology fanatics of the Italian Futurist movement vaunted in the early twentieth century.[276]

Without technology, people in industrialized countries would feel naked. Without our tools and machines, from microprocessors to gigantic coal excavators, we would not have been in a position in the first place to transform the geological, biological and chemical character of earth that is culminating in the advent of a new geological epoch. An extraterrestrial that landed here today would perhaps even see our technology as earth's primary characteristic.

People are technophiles by nature. Three million years ago, primeval

274. Igal Zeifman, Bot traffic is up to 68,5% of all website traffic, www.incapsula.com.

275. Koert van Meensvort and Hendrik-Jan Grievink, *Next Nature*, New York: Actar, 2012.

276. Lawrence Rainey et al., *Futurism—an anthology*, New Haven, CT: Yale University Press, 2009.

man fashioned the first biface tool, manufactured with merely five strokes. It took until 1.6 million years ago for humans to develop more complex tools that required twenty-five strokes. A quarter of a million years ago, instrument tips were developed that required one hundred strokes. Thirty thousand years ago, in an explosion of creativity, blades, arrowheads and stone axes appeared that required 250 strokes.[277] The wheel, metalwork and other key innovations followed. During the Song Dynasty in China, which lasted from 960 to 1279, coal was already being mined, and from 1498 on, crude oil was excavated for healing purposes in places like Pechelbronn.[278] Then in eighteenth- and nineteenth-century Europe, the combination of coal resources, a functioning patent system, early venture capital and capitalism itself resulted in the steam engine. An inventor called Denis Papin mounted one in the modern-day Bergpark in Wilhelmshöhe in Kassel, Germany as early as 1706. But it was only the successive work of Thomas Newcomen and James Watt that triggered the Industrial Revolution. The first commercial steam engines were used to pump water out of mines to avoid flooding.[279] Since water-free mines allowed even greater quantities of raw material to be extracted, a self-accelerating development set in: fossil fuels drove machines, and machines excavated new fossil fuels.

When Ludwig Meyn began to mine bituminous (or tar) sands near Hamburg, Germany in 1856, and Edwin Drake tapped the first big oil sources in Pennsylvania in 1859, the ground was prepared for today's hydrocarbon-driven world: A century and a half later, cars worldwide are run on fossil fuels. Chemists have released millions of new carbon compounds into the atmosphere.[280, 281] Except for being boxed in metal, steam engines and combustion engines have spread like wildfire across the world. Automation and production lines have given humans an entirely new role in earth's material cycles as we are now able to manufacture objects in

277. Numbers from the permanent exhibition of the Jena Phyletic Museum.

278. Daniel Yergin, *The Prize—the epic quest for oil, money and power,* New York: Free Press, 2008.

279. Akos Paulinyi and Ulrich Troitzsch, *Mechanisierung und Maschinisierung, Propyläen Technik Geschichte 1600 bis 1840,* Berlin: Propyläen, 1997.

280. Siegfried Zimmermann, "Ludwig Meyn und die Entwicklung der Erdölindustrie bei Heide in Holstein," Dissertation, Hamburg, 1966.

281. Daniel Yergin, *The Prize—the epic quest for oil, money and power,* New York: Free Press, 2008.

almost limitless quantities. Machines have turned us into earth-shaking transformers: in 2008, humans moved 80 billion metric tons of material, more than ten tons per capita.[282]

A collective "global factory" is growing, one that transforms stone, living organisms, and fuel into new technological ensembles on a grand scale. Chemistry, engineering, automation, communication, information processing, mobility, space travel, energy production—all these areas have gone through explosive developments in a matter of decades. Since the advent of the Internet, geographical borders have become virtually obsolete. Every person with Internet access has the public knowledge of the world at his or her fingertips; everyone can learn from and cooperate with everyone else. It has become commonplace for large teams, scattered across the globe in the most diverse of locations, to innovate together.

Technological historians rightly rack their brains over why all this took so long to start, why our technological Cambrian moment didn't happen earlier in human history. But, since about 1945, technological progress has been accelerating so fast that historians can hardly keep up with it.

Technology is like a jungle gym that grows along with our intellect. Because this jungle gym is so huge, and we have already reached such dizzying heights with mechanized civilization, the further spread of the technosphere seems almost inevitable. It is probably more accurate to speak of a co-evolution between humans and machines—a mutual dependency. People can no longer live without technology, and technology, at least for now, is dependent on humans. It would take a series of catastrophes for technological advance to stop or slow down.

The Anthropocene has grown out of human wishes and desires, plans and dreams, abilities and inabilities. But the environmental scientist Peter Haff sees a growing momentum in the technosphere, a tendency in technology toward a proliferation that obeys its own laws and uses people for its own ends: "Technology defends its mode of operation primarily by offering incentives such as abundant food, medicines, instant communication channels and other desiderata that bind, or even addict, humans to the system that produces them."[283] Seen from this perspective, it remains

282. Thomas Wiedmann et al., "The material footprint of nations," published online before print. doi: 10.1073/pnas.1220362110, *PNAS* September 3, 2013.

283. Peter K. Haff, "Technology as a geological phenomenon: implications for human

unclear as to whether humans will really remain the driving force behind this development. There is already talk about an artificial intelligence emerging out of Google's global algorithmic learning machine.[284]

How humans will influence forthcoming technological developments is therefore, all the more important. When we realize that the first web browser was only developed in 1990, and that all the web's functions have been developed in a very short space of time, we can only guess at what will be possible when the Internet—or whatever it will be called in the future and whatever it will be by then—has existed for 200 or even 2,000 years. It is quite likely that what we are currently going through are the fledgling stages of technological history—which is why we still have a disproportionately strong influence on the direction it will take from now on. Current technologies may function more or less well from the point of view of a network administrator, or the owner of a fully automatized factory, or the consumers of electronic goods, or drivers of cars. Many of the things that we deal with nowadays seem to "work" by some fantastic magic. But our current technologies fail to function in one crucial area. The technosphere has become its own global system like the global water cycle when it "appropriates Earth resources, including energy, mass and information, for its own uses on a large scale."[285] But in contrast to the water cycle, which constantly reuses its components, the technosphere disposes of its waste products into the biosphere: "Unlike earlier earth paradigms, which recycle most of their waste products, the technosphere does little recycling."[286] A large portion of what we refer to as "environmental problems" these days—climate change, electronic scrap, plastic pollution—derives from the fact that the technosphere is terrible at preventing waste and recycling its components in the way that leaves do when they fall from trees, or raindrops when falling from the sky.

The reason for this inability is obvious. Up until 250 years ago, something like waste hardly even existed: the majority of material cycles in

well-being," in: C.N. Waters et al. (eds.), "A stratigraphic basis for the Anthropocene," Geological Society, London, Special Publications no. 395, first published 25 October, 2013; doi:10.1144/SP395.1.

284. Christian Schwägerl, *Die analoge Revolution*, Munich: Riemann-Verlag, 2014.

285. Peter K. Haff, "Technology as a geological phenomenon: implications for human well-being," op. cit., see footnote 281.

286. Peter K. Haff, "Technology as a geological phenomenon: implications for human well-being," op. cit., see footnote 281.

operation were closed. And when waste was created on a large scale by industrialization, people were comforted by the trope of the "great boundless nature out there," with its unlimited resources that could be drawn upon to absorb and hide all our garbage.

Today we still think that technology is something that exists separately from nature. But this is not true. Technology comes from earth's crust; it is a fusion of intellect and geology. For the 1.75 billion cell phones alone that were sold across the globe in 2012, 300 metric tons of silver, 29 tons of gold and 11,000 tons of copper had to be excavated, leaving behind vast waste dumps.[287, 288] In order to run technical apparati, humans mine 32 million barrels of crude oil a year, 7.2 billion tons of coal, 3,400 billion cubic meters of natural gas, three billion tons of iron ores, 17 million tons of copper and many millions of tons of other ores and chemicals.[289] In computers and fiber-optic cables there are enormous quantities of silicon, the most common element in earth's crust. Machines, technical infrastructure and their products have already formed a new geological reality. Accordingly, a good proportion of our communication runs tens of thousands of miles through fiber-optic cables laid deep in the ocean, creating, in the words of Australian zoologist Tim Flannery, a "mammalian super-organism." Our smartphones are in fact elaborately processed rocks. They bind forty different elements from the periodic table, from tin to cobalt to palladium. Our cars, if lined up one after the other, would already cover an area the size of South Korea.

Once they have been used, these materials are still disposed of Holocene style—which, in most cases, means that they are simply dumped somewhere. Once technology turns to waste, it forms a new kind of geology: electronic scrap ends up in landfills in Asia and Africa.[290] Certain materials like rare earth elements (REE) that were formerly concentrated in very few places on earth, are suddenly turning up as "spice elements" all over the world because they are reaching the furthest flung corners of the

287. Sales figures source: http://www.gartner.com/newsroom/id/2335616.

288. http://www.unep.org/pdf/pressreleases/E-waste_publication_screen_finalversion -sml.pdf, p.8.

289. International Energy Agency, Paris; http://minerals.usgs.gov/minerals/pubs /commodity/.

290. See World Map project material on www.step-initiative.org.

planet via electronic products. Anthropogenic geologists in the future will be able to register these long-lasting changes.

As things stand, almost sixty million tons of electronic scrap accrues each year, seven kilos per capita on earth on average. But consumption is unevenly distributed. Billions consume only very little, while Americans average at least thirty kilos of electronic waste per year. This waste contains much more gold and other precious items than do ores from mines—materials valued to the tune of over forty-five billion dollars in total.[291] Currently, only a fraction of this electronic scrap (less than ten per cent) passes through the hands of recycling experts. Although tough geopolitical struggles are fought for REE, the "vitamins of information technology," thousands of tons of these end up in the garbage every year; there is no strategic recycling, according to the US Geological Survey.[292]

In an economic system in which wastefulness is defined as financial profit, the most attractive method in the short term is to excavate compact raw material deposits from the depths of mines using fossil energy and then scatter them in the form of electronic scrap all over the world. The obvious alternative of creating closed material cycles is employed on only a comparatively very small scale.

The result is that entire regions in China, India and along the West African coast are filling up with electronic wastes. People living there risk being poisoned as they retrieve metals from computers and cables over open fires. Day by day, millions of people buy new flat screens, new laptops, iPads or iPhones and throw away old ones. This results in a fine layer of metal and plastic spreading worldwide.

The present inability of the technosphere to recycle its materials can be illustrated in far greater dimensions: around the year 2000, the average urbanite produced one kilogram of mixed waste per day. A line of trucks of over 2,500 kilometers (1,500 miles) long was necessary to cart off one daily load. If current trends continue, the quantity of waste from cities will double by 2025 and humanity's entire waste load will grow by the year

291. Mathias Schlüp et al., *Recycling—from e-Waste to resources*, Berlin: UNEP 2009.
292. Goonan, T.G., 2011, "Rare earth elements—End use and recyclability," U.S. Geological Survey Scientific Investigations Report 2011–5094, available only at http://pubs.usgs.gov/sir/2011/5094/ and EPA, "Rare Earth Elements: A Review of Production, Processing, Recycling, and Associated Environmental Issues," EPA/600/R-12/572, December 2012.

2100 to 11 million tons a day;[293] this too will leave an infinite number of new anthropogenic geological traces.

It's great that many people today separate their waste in containers of different colors. Some of this material is actually recycled but the most important change hasn't happened yet: Product materials and design are *still* not based on the concept of reuse. In fact, quite the opposite is true: entire development departments are busy embedding predetermined breaking points into their products so that demand remains constant and profits continue to increase. Many products are not even reparable, or the cost of repair is kept so high, that it is cheaper to buy a new product.

Plastics best illustrate the extent to which the technosphere is removed from recycling. This group of substances, first developed in 1907 in the form of Bakelite, now symbolizes modern convenience and vital necessity as no other material does. Using plastic saves lives in medicine and in food hygiene. It allows a variety of different forms and functions for products in all areas of life. Plastic is extremely versatile and can be used for very environmentally friendly purposes, for example as a light construction material. But as a waste product, it stays in the environment for a very long time—up to centuries—where it only decomposes little by little, releasing its degradation products as it does.[294] Scientists have found astonishing new communities of microbes in the "plastisphere," many of which are carcinogenic.[295] But some plastic, according to the geologist Jan Zalasiewicz, will remain behind as a new type of fossil for a very long time. In 2014, scientists described a new type of "plastic rock" which they say will persist for a very long time.[296]

Factories all over the world produce more than 250 to 300 million tons of plastic a year from fossil resources. That is enough to wrap the entire surface of the United States once around in cling film,[297] an idea that might appeal to Christo, the artist who wraps entire islands or buildings

293. Daniel Hornweg et al., Environment: "Waste production must peak this century," *Nature*, 30 October 2013.

294. Susan Freinkel, *Plastic—a Toxic Love Story*, Boston: Houghton Mifflin, 2011.

295. Erik R. Zettler et al., "Life in the "Plastisphere": Microbial Communities on Plastic Marine Debris," *Environmental Science & Technology*, vol. 47, no. 13 (2013): 7137–7146.

296. Patricia Corcoran et al., An anthropogenic marker horizon in the future rock record, *GSA Today*, vol. 24, no. 6, (2014).

297. C. Rochman et al., "Classify plastic waste as hazardous," *Nature*, vol. 494, (February 14, 2013): 169–171.

in fabric. Across the globe, more plastic will probably be created in the next ten years alone than in the past 100 years.[298, 299] A large portion of this production will rapidly end up as waste.[300]

In highly developed countries, colossal quantities of plastic waste end up in landfills; in emerging nations, plastic waste has become a dominant part of the environment. Rivers transport plastic bottles, Styrofoam scraps, synthetic containers, bags, packaging and much else besides into the world's oceans.[301] Scientists keep finding new forms of "plastic plankton" in seawater.

In the North Pacific Ocean, ostensibly the deserted half of the planet, a colossal collection of debris from the confluences of Asia, America and garbage from ships has amassed to become the Great Pacific Garbage Patch. On some beaches in Hawaii, plastic particles outnumber grains of sand.[302] In the South Atlantic, scientists found more than 10,000 pieces of plastic in one hectare of water. In Kenya, there is a "flip-flop" coast onto which beach sandals from the Maldives and Seychelles are washed up by the ton.[303]

Hundreds of thousands of whales, fish, birds and turtles die agonizing deaths when they swallow these plastic parts. Sperm whales mistake plastic bags for the jellyfish on which they feed and bite marks in pieces of plastic reveal that many birds also mistake these bags for food.[304] Their bellies fill up with trash until they eventually die. Pieces of plastic were

298. David Barnes et al., "Accumulation and fragmentation of plastic debris in global environments," *Philosophical Transactions of the Royal Society*, vol. 364, no. 1526 (27 July 2009): 1985–1998.

299. Richard Thompson et al., "Our plastic age," *Philosophical Transactions of the Royal Society*, vol. 364, no. 1526 (27 July 2009): 1973–1976 as well as "Plastics—the Facts 2010, An analysis of European plastics production, demand and recovery for 2009," Brussels: Association of Plastic Manufacturers in Europe, 2010.

300. EPA, see http://www.epa.gov/osw/conserve/materials/plastics.htm.

301. Richard Thompson et al., "Lost at sea: where is all the plastic," *Science*, vol. 304, no. 5672, (7 May 2004): 838.

302. Tracy L. McMullen, Karla J. McDermid, "Quantitative analysis of small-plastic debris on beaches in the Hawaiian archipelago," *Marine Pollution Bulletin,* vol. 48, issues 7–8, (April 2004): 790-794.

303. Sarah Wachter, "Recycling Discarded Flip-Flops," *New York Times*, (8 Dec 2009).

304. http://www.theguardian.com/world/2013/mar/08/spain-sperm-whale-death-swallowed-plastic.

also found inside an albatross stomach, bearing a serial number that could be traced back to an airplane that had been shot down sixty years previously, almost 10,000 kilometers away.[305] Fish, whales and turtles become ensnared in old fishnets and plastic parts that float around in the ocean. The remote Pacific island of Midway is one of the many settings of this anthropogenic drama. It is where photographer Chris Jordan has documented how plastic from all over the world is washed ashore and then eaten by birds that die slowly and painfully as a result.[306]

Our penetration of nature through our wastefulness—through an over-flow of goods in the literal sense—is happening not just with larger objects but with much finer, invisible substances as well. The diversity of synthetic chemicals is constantly increasing. Ten million different chemicals, the majority of which were synthetically manufactured, were on register with the American Chemical Society in 1990. Toward the end of 2009, the Canadian pharmaceutical company, Chlorion, added the fifty-millionth substance to the list. Within that year, novel chemical substances were either isolated or being synthesized at the rate of one every two and a half seconds.[307]

While useful, life-saving, beautiful and wonderful objects have been created through modern chemistry, manufacturing practice is still marked by the spirit of the Holocene, featuring a "great outdoors" into which substances can vanish.

Poisonous mercury accumulates in fish because this heavy metal is used to free gold from rocks for jewelry, central banks and electronic equipment. Fish also absorb pharmaceuticals and chemicals used in personal hygiene.[308] In remote swamps, there are frogs that have undergone sex changes or become sterile from the effects of pesticides, contracep-

305. Kenneth Weiss, "Plague of plastic chokes the seas," *Los Angeles Times*, (2 August 2006).

306. Chris Jordan, Midway Project: see http://www.youtube.com/watch?v=PLkTTJ W4xZs.

307. See website of the American Chemical Society at: http://www.cas.org/newsevents /releases/50millionth090809.html.

308. Kevin Chambliss et al., "Occurrence of pharmaceuticals and personal care products in fish: Results of a national pilot study in the United States," *Environmental Toxicology and Chemistry*, vol. 28, no. 12, (January 2010): 2587–2597.

tives, plasticizers and flame retardants.[309, 310] Out of the sewage pipes of big Western cities, residues of medicines pass via urine into the water along with small pellets from shower gels that turn up in the bellies of fish. And very soon there will be bactericide nanoparticles—the latest rage among hygiene fanatics wandering along the food chain through the living world, operating invisibly.

Factories in Europe and America have become noticeably cleaner over the past few decades but often, this has only happened because production was transferred to Asia. In China in particular, pollution of water, air and soil has reached such critical levels that the health of the people is directly endangered.[311] While teenagers in Europe and the US hunt for cheap clothes, they ignore the abysmal treatment of workers in factories and the pollution that make the low prices possible. Fashion trends in Europe and the US can be recognized by the color the rivers run in Bangladesh.[312]

In the Anthropocene, it is no longer enough to see the usefulness in products such as electronic goods, plastic and chemicals: in the "invironment," everything comes back a million times over. The fact that doctors have found more than 200 potentially poisonous chemicals in the umbilical cord blood of newborn babies is just one among thousands of anthropogenic effects.[313, 314] The frequent smog in Asia is a kind of precursor to the cheap products that end up on shelves in the Western world but exposes

309. See Krista McCoy et al., "Agriculture Alters Gonadal Form and Function in the Toad Bufo marinus," *Environmental Health Perspectives*, vol. 116, no. 11, (November 2008): 1526–1532.

310. Matthew Milnes et al., "Contaminant-induced feminization and demasculinization of nonmammalian vertebrate males in aquatic environments," *Environmental Research*, vol. 100, no. 1, (January 2006): 3–17.

311. Jonathan Ansfield and Keith Bradsher, "China Report Shows More Pollution in Waterways," *New York Times*, (February 9, 2010).

312. Jim Yardley, "Bangladesh Pollution, Told in Colors and Smells," *New York Times*, (July 14, 2013): http://www.nytimes.com/2013/07/15/world/asia/bangladesh-pollution-told-in-colors-and-smells.html.

313. Environmental Working Group, "Pollution in people—cord blood contaminants in minority newborns," Washington, DC, 2009.

314. U.S. Centers for Disease Control and Prevention, "Fourth National Report on Human Exposure to Environmental Chemicals," 2009, and http://www.cdc.gov/exposurereport/pdf/NER_Chemical_List.pdf.

hundreds of millions of people to health hazards.[315] Radioactive effluent from the Fukushima disaster ends up in edible fish on the American west coast.[316] The Anthropocene is only partly what we do; the other part is what happens to us as a consequence of what we do.

The majority of the problems that I have described here are not problems intrinsic to technology: they result from the fact that, in our economy, both resources and dumps are still perceived as inexhaustible. But the geologist Peter Haff believes that we have reached a critical point: "The future of the technosphere as a paradigm rather than just an episode in earth history is contingent upon the emergence of effective recycling mechanisms."[317]

At the beginning of the Anthropocene, the issue is whether the extremely young technosphere overruns and conquers the biosphere or, by its example, learns how to constantly recycle material. This requires the important virtues of frugality and moderation in consumption because only by consuming less in the short term will we cause less waste. It is also important to improve the efficiency of equipment. This has its limits, however: when an oversized SUV uses five per cent less fuel and will be replaced by a new one within a short time, nothing is won. Besides, improvements in efficiency are often counterbalanced by more consumption: people just buy bigger models of energy-saving refrigerators, an outcome known as "rebound effect." To simply count on efficiency means, as Michael Braungart puts it, "to destroy the planet more slowly and more thoroughly." This follows the logic that, using the wrong technology, only wrong things can happen in the long term no matter how efficiently or frugally they are implemented.[318]

315. Joanna Foster, "China's first smog clinic opens its doors," *Think Progress*, (December 18, 2013): http://thinkprogress.org/climate/2013/12/18/3083901/chinas-smog-clinic/.

316. Vincent Rossi et al, "Multi-decadal projections of surface and interior pathways of the Fukushima Cesium-137 radioactive plume," *Deep Sea Research I*, vol. 80, (June 2013): 37–46, http://web.maths.unsw.edu.au/~matthew/Rossi_et_al_DSR_2013.pdf.

317. Peter K. Haff, "Technology as a geological phenomenon: implications for human well-being," op. cit., see footnote 281.

318. Michael Braungart and William McDonough, *Cradle to Cradle*, Vintage, 2009; and Michael Braungart and William McDonough, *The Upcycle—Beyond Sustainability, Designing for Abundance*, North Point Press, 2013.

What is actually needed is a more profound change in technology itself. Just like agriculture and cities, technology also will have to function as "new nature" in the Anthropocene. Constructing machines that function in the same way as living organisms has exerted a great fascination on humanity since time immemorial. This notion is exemplified by Leonardo da Vinci's early sixteenth century flying machines, Jacques de Vaucanson's eighteenth-century mechanical duck, and countless modern-day projects like humanoid robots and bird-like drones.[319, 320, 321] There are certainly nightmarish scenarios at the interface between technology and nature. One of these is a plan by DARPA, the US military's Defense Advanced Research Projects Agency, to design hard drives that dissolve on command if they fall into enemy hands on the battlefield.[322] If these devices were to lead to toxic waste entering groundwater, the environmental fallout of war would become even greater. On the other hand, if DARPA's research were to lead to the development of compostable machines, it would be a genuine breakthrough.

Future technology has to consist of machines, materials and molecules that adapt to the biologic cycles of earth instead of perturbing them, and they have to enrich earth with life-enhancing stimuli instead of discharging poisons.

What is needed, therefore, is a different, new "nature of technology," an evolution whereby technology adapts to its environment. The more scientists reveal nature's inner mechanisms, the more primitive today's technology looks in comparison. Brian Arthur, the American technology theorist, sees natural mechanisms, materials and designs as role models for future technology: "Living things give us a glimpse of how far technology has yet to go. No engineering technology is remotely as compli-

---

319. Gaby Wood, *Edison's Eve*, New York: Anchor Books, 2002.

320. Allen McDuffee, "Army Scores a Super-Stealthy Drone That Looks Like a Bird," *Wired*, (November 27, 2013): http://www.wired.com/dangerroom/2013/11/army-maveric -microdrone/.

321. Illah Nourbaksh, "Google's Robot Army," *New Yorker*, (December 18, 2013): http:// www.newyorker.com/online/blogs/elements/2013/12/when-robots-become-smog.html.

322. Spencer Ackerman, "DARPA wants to create dissolvable spy hardware," *Wired*, (January 29, 2013).

cated in its workings as the cell."[323] Arthur sees the future of technology as "self-healing, self-configuring, cognitive," and "organic."

Seen from the perspective of a future bio-technosphere, today's wasteful machines appear to be rudimentary organisms with outdated cycles in urgent need of improvement. Fuel-guzzling Porsches or SUVs, coal-fired power plants and persistent plastic seem to be old-fashioned leftovers from the Holocene, about as impressive as horse carriages and typewriters. Cars of the future would either decompose into material that boosts the environment or give way to other, networked kinds of transport. The guiding principle in this process might be called *bioadaptation*: using nature as a source to "breed" machines.

A rapid transition from fossil and nuclear, that is, from degenerative energy sources to renewable energy sources emerges as an absolute priority in this perspective, as do large-scale programs to extract carbon from the atmosphere in a biological way, with the help of restored woods and moors. According to what scientists have stipulated, if we are to reduce carbon emissions by eighty per cent by 2050 and, if only an additional 269 billion tons of carbon dioxide is allowed to end up in the atmosphere in this century, then our first priority is to put all technology to this use.[324]

This could be the starting point of a far greater transformation to a completely regenerative economy, a closed cycle economy, in which the term waste no longer means anything. If products were built so that they could be turned back into raw materials for the next generation they would be easier to repair and upgrade than they are today. These products could be made from plant-based materials optimized by biotechnology. Or they could be synthetic and feature organic characteristics created through biomimicry. Innovative recycling plants could be fed with entire scrap heaps from the past to be processed into new raw materials through urban mining.[325] Plastic recycling would become a source of raw material,

323. Brian Arthur, *The Nature of Technology—what it is and how it evolves*, New York: Free Press, 2009.

324. Andrew Freedman, "IPCC Report Contains 'Grave' Carbon Budget Message," *Climate Central*, (October 4, 2013): http://www.climatecentral.org/news/ipcc-climate-change -report-contains-grave-carbon-budget-message-16569.

325. Gemima Harvey, "The Potential of Urban Mining," *The Diplomat*, (November 19, 2013): http://thediplomat.com/2013/11/the-potential-of-urban-mining/.

making it superfluous to drill new oil platforms in the deep sea or in the Arctic. New chemical substances would absorb toxic substances and render them harmless.

The bioadaptive imagination of scientists is only just gathering momentum:

- Compostable cars; synthetic materials that turn into nutrients when they dissolve
- machines made of organic material that could easily be recycled
- electrodes that work with endogenous substances
- colorants following the role model of butterflies
- substitute plastic made of insect protein
- biodegradable electronics
- robots that feed off plastic waste
- nanomagnetic designer particles that extract phosphorus and other critically important elements from wastewater
- buildings inspired by deep-sea sponges
- power plants that imitate photosynthesis
- bacteria that produce fuel and construction materials
- signal transmission in silk threads.[326, 327, 328, 329]

The transformation could go even deeper: genetic algorithms in the future could enable much more complex calculation processes than digital ones; biological nanomaterials could significantly reduce the necessity to use metal and finally, DNA—the very stuff of which life is made—could prove to be superior in information storage, making DNA computers a reality.[330]

326. Christopher J. Bettinger, "Biologically derived melanin electrodes in aqueous sodium-ion energy storage devices," *PNAS*, vol. 110, no. 52 (December 24, 2013): 20912–20917.

327. Philip Ball, "Silk lasers: 'Edible' electronics move closer," *BBC Future*, (September 14, 2012): http://www.bbc.com/future/story/20120907-lasers-made-from-silk.

328. Javier G. Fernandez and Donald E. Ingber, "Unexpected Strength and Toughness in Chitosan-Fibroin Laminates Inspired by Insect Cuticle," *Advanced Materials*, vol. 24, issue 4, (January 2012): 480–484.

329. Peter Fratzl et al., "Micromechanical properties of biological silica in skeletons of deep-sea sponges," *Journal of Material Research*, vol. 21, no. 8 (August 2006).

330. Nick Goldman et al., "Towards practical, high-capacity, low-maintenance information storage in synthesized DNA," *Nature*, vol. 494, (February 7, 2013): 77–80.

An inkling of what is possible can be found in some wonderful creatures called slime molds. Their name might cause revulsion in some but *myxomycetes*, as they are called in Latin, are some of the most fascinating organisms on earth. These unicellular entities cannot easily be categorized as animals or plants, and they are also not fungi. They can sometimes be spotted in woodland areas in the form of yellow slime on tree trunks. When they are visible to the naked eye they are at a stage known as plasmodia, where single cells have united to form multiple nuclei that start to move and can collectively react to environmental stimuli such as light or nutrients. The mass of single cell organisms is then able to creep to places where there is a food supply, often spreading over a larger area in the form of a net.

This kind of slime mold has shown scientists what bioengineering assets have yet to be exploited; or to be more precise, they show us that the story of technology has only just begun. For urban architects, it is often a difficult task to connect many places in a city quickly and at low cost: complicated mathematical puzzles lurk behind solving these problems. Other network administrators, such as managers of electricity supplies, financial systems and transport chains, have similar problems. A spectacular experiment by Japanese and British scientists using a type of slime mold (*Physarum polycephalum*) resulted in the design of a functioning railway network for the metropolitan area of Tokyo without either centralized control or planning. For the experiment, the scientists laid oat flakes on a glass surface to imitate the way in which locations around the Japanese capital were distributed. They then let the slime mold grow onto the model of the city. With astonishment, they observed the way in which new connections grew before their eyes: the slime molds had detected the shortest, and most efficient routes that would have normally required great computing power: "The networks showed characteristics similar to those of the rail network in terms of cost, transport efficiency, and fault tolerance," but "the *Physarum* networks self-organized without centralized control or explicit global information."[331] In the future, people might be able to reach their destinations at significantly higher speeds if slime molds are

331. Atsushi Tero et al., "Rules for biologically inspired adaptive network design," *Science*, vol. 327, no. 5964, (22 January 2010): 439–442.

involved in traffic planning. New navigation systems could also be run using the intelligence of slime molds: "powered by iMyx." Such experiments could also provide models for an intelligent and environmentally friendly linked-up traffic system, in which a mixture of electric bikes, car sharing services, comfortable buses and closely spaced trains make individual cars look like a crude waste of resources and urban space.

Biotechnology of this kind is best suited to creating the next big wave of innovation whereby the technosphere learns from the biosphere. And this wave had better come soon because the wave of old technology is still spreading out like a tsunami in increasing number of coal-fired power plants, more cars burning fossil fuels, and more and more plastics, while digital media is being used to fan consumerism. Internet platforms accelerate consumerism and material turnover; powerful algorithms construct a "digital twin" from the data and patterns of our online existences that constantly accompanies us and encourages us to indulge in more consumerism.[332, 333] The world of software apps gives us the impression that we can restructure our everyday lives; however, this simply masks new forms of governance by machines, especially as far as concentration management is concerned—our most important and precious neuronal resource. The kinds of machines we produce and how we produce them transforms not only the purely physical surface of the planet but also the human psyche. The kinds of machines that are created will strongly influence how the Anthropocene plays out.

Bioadaptation could be a new paradigm of technological development. That would put the work of the great natural historians like Maria Sibylla Merian, Carl von Linné, Alexander von Humboldt, Ernst Mayr and Edward O. Wilson in a completely different light, transforming them from their status as explorers of a vanishing world to the avant-garde of future technology. Humboldt, especially, was a pioneer in the technological development on which our natural sciences are based. But more than

332. Chandran Nair, "Five dangers the internet poses to a sustainable world," *Global Institute for Tomorrow*, (August 9, 2013): http://www.global-inst.com/ideas-for-tomorrow/2013/five-dangers-the-internet-poses-to-a-sustainable-world.html.

333. The phrase "Digitale Zwilling" ("virtual twin") was coined by the German President Joachim Gauck during the National Security Agencies electronic eavesdropping scandal, October 2013: http://www.bundespraesident.de/SharedDocs/Reden/EN/Joachim Gauck/Reden/2013/131003-Day-of-German-Unity.html.

150 years after his death, we have only just begun researching the natural riches of the planet that would be vital to bioadaptive technology.

Around 2,300 years ago Theophrastus, designated successor to Aristotle, began making a scientific catalogue of living organisms in which he described approximately 500 plant species.[334] Since then, more than 300,000 known plant species, 5,500 mammals, nearly 10,000 bird species, and 1 million insect species have been identified. Around 1.9 million extant species have been described to date, but the complete number of species, according to scientists' estimates, is in the region of 9 to 11 million, not including bacteria.[335]

In the canopies of the rainforests, in the bio-niches of the savannah, in the depths of the ocean and in the underground lakes of the Antarctic, dwell many millions of species that no human has as yet consciously perceived. The entire knowledge of so-called primitive peoples, as well as expeditions by natural scientists undertaken from the nineteenth-century to the modern-day, provide only a very sketchy picture of life: everything that winds up in the natural history museums in Paris, London, Berlin and Washington gives us only a tiny cross-selection of reality.[336] We humans still have little idea with whom we share the planet. While most mammals and flowering plants are known, there are many other groups—insects, bacteria, viruses or algae—where our knowledge is rudimentary.

This lack of knowledge is mostly due to the fact that funding for the work of taxonomists and biodiversity experts leaves much to be desired. The academic "factories" that turn out MBA graduates are often smart new buildings made of glass, but to find the departments of taxonomy and systematics, you typically have to head for the most dilapidated buildings on a university campus. There is a global network of biological treasure troves in which scientists keep a variety of cultivated plants. Especially in developing countries, these centers are often in deplorable condition. Even in some wealthy countries, gene banks are neglected, with the notable

334. Peter Tallack, *The Science Book*, London: Weidenfeld and Nicholson, 2006.

335. Arthur D. Chapman, "Numbers of living species in Australia and the world," Canberra: Australia Biodiversity Information Service, 2009) and Camilo Mora et al., "How Many Species Are There on Earth and in the Ocean?" *PLoS Biology*, vol. 9, no. 8, (August 2011).

336. Charles Godfray and Sandra Knapp, "Taxonomy for the 21st century," *Philosophical Transactions of the Royal Society*, London, vol. 359, no. 1444 (2004): 559–569.

exception of Norway's Svalbard Global Seed Vault. And there are very few sponsors for the classic type of expeditions that would be necessary to track down the 7 to 9 million species unknown to us.

Edward O. Wilson, the biodiversity scientist and one of its most prominent advocates, established that there are only about 6,000 taxonomists worldwide today who are trained and able to differentiate and describe species precisely.[337] There are far too few people with the expertise to render global natural riches for the purpose of technological development. Wilson puts the cost of a global database, in which all species would be successively documented, at 60 million dollars—money that is not available despite the fact that it is a drop in the ocean compared to the billion-dollar packages given to banks during the "financial crisis" that started in 2008.[338]

Why is our civilization not in a position to bring forth 60 to 70,000 biodiversity specialists and to financially support them so that we can expand our understanding of the world around us and start using it properly? The reason has to do with our current Holocene economy, in which resources are wasted and costs are passed on to the general public or to the future because that is the quickest way to make money. But if the economic value of nature became clear, its destruction would become officially uneconomical, and this step would throw today's entire economic system into question. Human communities that generate their own energy and recycle their raw materials, whose fundamental satisfaction is not based on continuous consumption and who take resources from sustainable diversity, are the nightmare of those who are truly powerful in today's economic system. This is why the annual global sum of 88 billion dollars spent on subsidies for renewable energy sources is turned into a scandal by public relation strategists for the fossil fuel industry whereas the 523 billion dollars spent on subsidies for fossil fuels is hardly mentioned.[339] And this is how it is possible for banks to be declared "system-relevant" and saved with astronomical sums of money whereas plant

337. Edward O. Wilson, "Taxonomy as a fundamental discipline," *Philosophical Transactions of the Royal Society London* B, vol. 359, no. 1444 (2004): 739.

338. Edward O. Wilson et al., "The barometer of life," *Science*, vol. 328, no. 5975, (9 April 2010): 177.

339. International Energy Agency, "World Energy Outlook 2012," Paris, 2012.

banks and other real sources of lasting wealth are simply rotting away in many places.

Yet the economic and ecological potential of bioadaptation is virtually infinite. The IT (Information Technology) revolution has already demonstrated what scientists and engineers can do. The physicist Richard Feynman expressed the basic concept of today's computer technology in his often-quoted lecture from December 1959: "There's plenty of room at the bottom."[340] He described how great encyclopedias, in fact "all information that all of mankind has ever recorded in books," can be stored on ever-smaller devices. It was if he were describing a new continent.

Feynman's ideas inspired physicists and engineers to delve into the world of micro and nanoscales. The lecture that he gave after dinner at a meeting of the American Physical Society was received at the time with amusement.[341] Later on it was seen as the germ of revolution. Most of what Feynman described back then as utopia is now reality. Under the right conditions—changed consumer behavior, changed subsidies, a political focus on research and development—today's technology can become bio-adaptive. One evening at dinner with friends in the future, we might be discussing how inconceivable it is that people in the primitive past used to burn crude oil or make disposable packaging.

A new technological revolution could transfer Feynman's prophecy of computers onto the analog world: "plenty of room at the bottom" could be applied to technology that is created from biological sources. Instead of today's Cartesian machines, for which nature is an automaton and pure raw material, tomorrow's world needs something "Humboldtian:" machines capable of interconnecting with their ecological and energetic environment—machines that think like planets.

Feynman was a good-humored prophet and indicated this direction in his 1959 lecture, saying: "Biology is not simply writing information; it is doing something about it. A biological system can be exceedingly small. Many of the cells are very tiny, but they are very active; they manufacture various substances; they walk around; they wiggle; and they do all kinds of marvelous things—all on a very small scale. Also, they store information.

340. See http://www.feynman.caltech.edu/plenty.html.
341. Engineering and Science, vol. LXXIII, S. 1, Winter 2010.

Consider the possibility that we too can make a thing very small which does what we want—that we can manufacture an object that maneuvers at that level!"

Nano-biotechnology could supply the services afforded today by crude oil, metals, plastic and cement, by using innovative high-tech natural materials. The spectrum of these technologies is broad: wafer-thin sheaths for buildings instead of heavy cement, biomimetic packaging instead of fossil plastic, 3D printers that constantly manufacture new products made of recyclable raw materials, cutting out long transport routes.

For this to happen, however, far-reaching political decisions are necessary that drive the further development of technology in the right direction.

- The massive subsidies worldwide for fossil energy have to disappear

- There has to be an international climate agreement *very soon* that significantly limits the quantity of carbon from technology that is permitted to enter the atmosphere

- Recycling laws have to have a clear policy goal of managing material in complete cycles

- Scientists who explore the biological richness of earth should receive appropriate budgets

- Spending on research for technologies inspired by biology has to increase

- Restoration costs for damage done to the services of the ecosystem must be paid for by those that cause it.

The American shale gas revolution is a setback for this kind of development. While it does make the United States less dependent on countries like Saudi Arabia, the fracking boom creates the illusion that the USA can continue to bank on fossil fuels, and decelerates the expansion of renewable energies. Further, fracking also makes energy so inexpensive that it does not seem worthwhile to invest in more efficient equipment and recyclable materials. Instead of adapting technology to the biosphere, fracking adapts the biosphere to technology.[342] There is no doubt that fracking will leave behind long-lasting, visible geological traces, not least of all due to the chemicals used to loosen fossil fuels from rock. But I don't think that people will be proud of these traces. At worst, they will be a symbol of humanity's failure in the early twenty-first century to slow down climate change.

The principle of bioadaptation is suited to giving technology a different character: from being an opponent of the earth system to a symbiont. That doesn't mean that a world in harmony with nature would result as bioadaptation alone doesn't guarantee gentler technology nor are the products of biological evolution intrinsically good or moral. When technology and life interweave, questions of design and control become even more complicated than they are today with classic technology. Applications of bioadaptation can be negative or hazardous. That has been shown with micro-drones in the form of insects or birds that have been developed for surveillance, or in the four-legged robots from Boston Dynamics that are being designed as futuristic beasts of burden for the US military.[343, 344, 345]

But without bioadaptation, the technosphere will soon be not just a foreign entity in the earth system: the earth as a whole will transform into a foreign entity. One of the characteristics of technology, so it seems, is that while there is a solution for every problem, every solution introduces a new problem. The question now is whether we try to solve old problems with old technology, as in the so-called shale gas revolution; or whether we are ready to move on to new problems with new technology.

342. This shows a sad anthropogenic landscape: http://www.flickr.com/photos/amy-myou/9431314171/.

343. John Markoff, "Google Adds to Its Menagerie of Robots," *New York Times*, (December 14, 2013): http://www.nytimes.com/2013/12/14/technology/google-adds-to-its-menagerie-of-robots.html.

344. The US Department of Defense is already actively developing hybrids made of insects and robots: https://www.fbo.gov/index?s=opportunity&mode=form&id=ec6d6847537a9220810f4282eeddaod2&tab=core&_cview=1. This is also interesting: Robo-Roach, https://backyardbrains.com/products/roboroach.

345. Allen McDuffee, "Army Scores a Super-Stealthy Drone That Looks Like a Bird," op. cit., see Footnote 318.

NINE Directing Evolution

THE INVITATION ARRIVED at my home on thin airmail paper, written in a delightfully old-fashioned hand. Not long afterwards, we rang the bell at a nondescript door on the second floor of an old people's home called Badger Terrace, in Bedford, Massachussets near Boston. The man who received us was tall, and his cheeks glowed fresh pink as he smiled warmly at his two visitors. Before us stood the most important evolutionary biologist after Charles Darwin, Ernst Mayr, for whom an entire library at Harvard University has been named. He was then 97 years old and graciously ushered us into the small two-bedroom apartment to which he had moved after the death of his wife. He showed us his two desks that were piled high with current scientific journals, a compendium of Melanesia's bird world that he had published with Jared Diamond, and his latest book: *What Evolution Is*. Even at a ripe old age, he was still extremely creative and productive, which is why my colleague Joachim and I were interviewing him for the newspaper *Frankfurter Allgemeine Zeitung*.[346]

In our three-hour conversation, Ernst Mayr took us on an exciting trip through his life and twentieth century biology. Drawn to bird-watching as an adolescent, he was sent on an expedition to the highlands of Papua, New Guinea, at the age of twenty-six by the Berlin Natural History Museum; this later landed him a job with the American Museum of Natural History in New York. His greatest achievements were not in the field of ornithology, however; they were in the much bigger field of evolutionary biology. In the forests of Papua, Mayr told us, he'd had a lot of time to

346. Joachim Müller-Jung and Christian Schwägerl, "Darwins Apostel," *Frankfurter Allgemeine Zeitung*, (12 March, 2002)

think, and it was there that ideas came thick and fast about how to synthesize Charles Darwin's theory of evolution with more recent scientific findings in the field of genetics. In 1942 he put these ideas to paper and laid the foundation of modern biology with his "modern synthesis" before further advancing this field as a professor at Harvard University.

We spoke to him about the likelihood of intelligent life forms on other planets, the ways in which biology differs from physics, and asked his opinions on genetic engineering and cloning. He gave long, profound and detailed answers to all our questions. Mayr told us so many things with such precision that, despite being significantly younger than him, the two of us were completely exhausted after three hours of discussion, whereas he, on the other hand, continued to gush with information. He seemed to be infused with inexhaustible energy and an almost childlike urge to learn new things and understand fundamental phenomena in biology. That's why, when I think back on that interview, I focus mostly on the question I *didn't* ask him, simply because I wasn't familiar with Paul Crutzen's idea at the time: "What will happen to evolution in the Anthropocene?"

This is one of the most exciting issues in the entire debate surrounding the geological epoch of humans: blind chance—the factor quite rightly singled out by Darwin and Mayr as central to evolution—is now steadily being complemented by a new global force. This is not to say that mutation and selection, the twin forces through which life forms change and new life forms develop, will no longer be present in the Anthropocene. A logger who wipes out a rainforest species does not purposely intervene in evolution—and perhaps not even a biotechnologist who alters the genome of a bacterium wishes to manipulate the course of earth's history. But millions of conscious human actions have accumulated into a new force that works in a way that is far less blind and unreflecting than chance, which governed the preceding 3.8 billion years of life history.

What will happen to life on earth in the future? Hypothetically, if scientists study the Anthropocene in one million years, what will they consider to be the defining geological evidence of our age? Will it be us humans because we will leave behind not just a few scattered fossils like the hominid Lucy, but instead billions of skeletons? Or life forms that we ourselves have procreated or resuscitated? Or the life forms that have vanished in a brief Sixth Wave of Extinction brought about by reckless exploitation through human hands? Or perhaps it will be varieties and species that will only evolve in reaction to the anthropogenic transformation of earth—the results of climate change, marine acidification or urbanization?

Whatever they may be, they will have evolved in a different way from present-day humans. We are witnessing the transition from blind evolution according to Darwin's laws, to a conscious evolution directed by the human mind. We human beings have long since evolved into a force that reproduces not only itself, but masses of new life forms. Life itself is going through a human-made bottleneck.

The earliest example of this is how the wolf slowly turned into the dog as a result of living in proximity to people; humans were a constant factor in deciding which animals reproduced and which didn't. Through selective breeding, 343 officially recognized dog breeds have descended from wolves—from the 6 inch high Chihuahua to the 4 foot tall Great Dane. Human tastes and needs have shaped these breeds and their behaviors, from aggressive pitbull terriers to attentive sheepdogs. Goats, cattle and sheep have all undergone a similar process. Scores of crossbreeds can be added—because, as far as dogs and other animals close to humans are concerned, humans strive for the *creation* of biodiversity rather than its destruction, and they do the same with plants.

The earliest farmers were also the first plant cultivators, albeit simple ones. From many wild grasses, they selected those varieties of cereal whose grain remained in the seed head when ripe instead of scattering. Some of the ripe plants were then kept as seed for the following year. The first cultivators noted which plants thrived best in which soil and weather conditions, and the strength of their resistance to various pests. Farmers became agents in the game of genetic modification, soil quality, and the struggle for existence. This facilitated the evolution of tens of thousands of cultivated crop varieties such as corn, rice and wheat, all long before the first endeavors in scientific crop cultivation with its experiments in cross cultures, or Gregor Mendel's in-vitro fertilization of plants in the nineteenth century. Nowadays, seed banks and fields all over the globe accommodate tens of thousands of plant varieties.

Floriculture is another area where humans induce biodiversity. We only have to think of the many thousands of cultivated orchid species— among them the Indonesian *Dendrobium* with its political namesakes as diverse as Gorbachev, Angela Merkel and Kim Il-sung—to appreciate the biological creativity of humans.[347]

However, evolution in the Anthropocene does not only arise through conscious intentions. It also occurs through chance, accidents, unintentional hybridization, the spread of species to new habitats or the creation of new life forms and behavior patterns in extreme, man-made biotopes. In Mexico, birds have begun to build their nests with cigarette butts because the nicotine wards off parasites—survival of the fittest here is aided by human waste.[348] Crows in Tokyo use clothes hangers to build their nests— perhaps in a million years, a future biologist will find such a nest and wonder what has been going on.[349] The wingspan of cliff swallows in the United States has shortened in the past few decades—directed by the kinetic force of cars. It took a while for researchers to fathom the reason: in the 1980s, cliff swallows began to use highway bridges as breeding grounds because the bridges resembled the birds' natural habitat. Automobiles were a huge factor in the selection process: birds that did not swerve out of the way in time died in collisions, often before they could reproduce. Over the years, the numbers of short-winged swallows increased, as it was easier for these birds to quickly swerve than for those with longer wings. Quite simply, the longer-winged birds were killed on the roads while the shorter-winged ones were able to use their new habitat as a breeding ground.[350] These new behavioral patterns and characteristics can lead to significant evolutionary consequences over time, even to the emergence of new species.

Another groundbreaking factor in future evolution is likely to be the increase of living creatures that are shaped by humans. Cat-lovers tend to believe that cats control people and not the other way round. But without human affection, feeding and regular costly trips to the vet, many of the

347. Brandlhuber: "Kim Jong Il, Kimilsungia, Pyongyangstudies IV," AdbK Nürnberg (2008), http://www.a42.org/fileadmin/_img/disko/disko_11.pdf.

348. Matt Kaplan, "City birds use cigarette butts to smoke out parasites," *Nature*, (December 5, 2012).

349. Yosuke Kashiwakura, Entry for National Geographic Photography Contest 2013: http://photography.nationalgeographic.com/photography/photo-contest/2013/entries/gallery/nature-winners/#/3.

350. Charles Brown and Mary Bomberger Brown, "Where has all the roadkill gone," *Current Biology*, (March 18, 2013): http://download.cell.com/current-biology/pdf /PIIS09609 82213001942.pdf?intermediate=true and http://www.humanesociety.org/ issues/pet_overpopulation/facts/pet_ownership_statistics.html.

100 million housecats in the USA would not exist, nor would the much greater number of half-feral strays. In the 1990s the population of domestic cats worldwide was estimated at 600 million, but the figure is likely to be much higher now. Cats have only spread across the planet due to humans. They have even reached earth's 179,000 islands where they kill huge numbers of birds and small mammals.[351] In the United States alone, cats kill between 1.4 to 3.7 billion birds and 6.9 to 20.7 billion mammals annually according to latest estimates."[352]

Killing and selection are ways in which human-cat relationships are changing our planet. But feline evolution doesn't stop there. At the beginning of the Anthropocene project at Berlin's Haus der Kulturen der Welt in 2013, the geologist Jan Zalasiewicz put the figures into context: "There are at least 100,000 domestic cats in the world for every tiger in the wild." He put forward the hypothesis that today's housecats could potentially dominate feline evolution rather than tigers, and that in the distant future, a new tiger-like creature might roam the earth whose ancestor was the domestic cat.

These are all huge influences on life and evolution but in comparison to what biologists are capable of these days, it just scratches the surface. In chemistry and biology laboratories, something much more fundamental is going on. The rediscovery of Gregor Mendel's manuscripts initiated the search for what it is exactly that determines heredity. As long ago as 1869, a strange substance was detected inside cells: deoxyribonucleic acid. In the early twentieth century, its chemical composition became known. In 1953, James Watson and Francis Crick delivered the model of the DNA as a double helix, which made it possible to both understand and alter the mechanism of heredity transmission. In 1972, scientists were able to create the first DNA that had ever been modified in a laboratory and followed this in 1973 with the first genetically modified mouse.

Humans had an impact on the genetic makeup of plants and animals for centuries but with the knowledge of DNA, genes and heredity,

---

351. Félix M. Medina, "A global review of the impacts of invasive cats on island endangered vertebrates," *Global Change Biology*, vol. 17, issue 11, (November 2011): 3503–3510.

352. Scott R. Loss et al., "The impact of free-ranging domestic cats on wildlife of the United States," *Nature Communications*, vol. 4, no. 1396 (29 January 2013).

the door to an entirely new dimension was flung open. At the end of the 1970s, scientists embarked on a mammoth enterprise to decode the genetic sequence of living organisms, building block by building block. Since the 1980s, in laboratories the world over, innumerable new life forms have been created in which genetic traits from the most diverse origins have been fused. In the meantime, some of these life forms have become the "norm" in agriculture, especially in North and South America. Among these are corn and soy whose resistance to pests has been boosted with the help of bacterial genes. There are fish that glow with the help of algae genes; apple trees that have butterfly genes implanted to help defend them against aggressors; genetically altered pigs whose urine has been cleansed of environmental contaminants; and a million mice strains whose diseases are supposed to simulate those of humans.

Only a small proportion of these living creatures would be capable of surviving in the wild—but what does that even mean in a world in which zoos have long since guaranteed the survival of many species, and where laboratories have become new habitats? Enormous laboratory-zoos of genetically modified animals have sprung up where experiments are carried out to determine the effects on cells of removing or adding this or that gene. Biomedical labs represent a new type of habitat in the Anthropocene era in which different laws operate than in the former wild. In nature, only the strongest and fittest survive; this contrasts with the sterile atmosphere of animal testing institutes, where researchers breed and reproduce the sick, weak and susceptible in order to study human illnesses. A slightly confused Sumatran tiger will soon fall prey to poachers, but a very confused mouse is cause for scientists to celebrate a deeper understanding of Alzheimer's. Whether these mice reproduce in the long run, and under what circumstances, remains completely open: a gigantic field test would be unleashed if animal activists freed such mice from biomedical centers.

Biotechnology has created a new situation: genetic information can jump across species much more freely. Some viruses and bacteria have been doing that for a long time, but scientists are now able to swap genes back and forth between bacteria, animals, fungi and plants in a conscious fashion. As Freeman Dyson has declared, a similarly free flow was possible only during the period before species' boundaries came into existence in the early phase of bacterial evolution. Organisms exchanged their genes more or less freely then too, but not according to the will of one single vertebrate.[353]

Scientific institutions tend to hide these new man-made breeds in order to avoid controversy. Societies want to benefit from meat production, medical research and animal testing but shy away from looking at the implications and the suffering involved. Not many people want to know what goes on behind the gates of test labs and agricultural factories for fear of the unpleasant facts they might find out and the difficult debates that might ensue. This can be illustrated using an animal very close to us that has been central to the ascent of humanity during the Anthropocene: the cow.

There are 1.4 billion cows worldwide; which is a ratio of about six people to one animal. The everyday lives of billions of people are linked to cattle—they eat meat, drink milk and live in landscapes shaped by cattle agriculture. People are greatly indebted to the cow as it was only after its domestication 6,000 years ago in the Middle East and India that the civilizational leap of humans was possible. All of a sudden, people no longer had to scratch away at the ground; they were able to cultivate larger areas of land. This freed up time in which they could invent and think. When the first people began to tolerate lactose and also drink milk, a very close relationship to the cow began.[354]

Alpine pasture farming, which allowed people to settle in high mountain areas and fill entire regions with humans, would have been inconceivable without cows. On stretches of land with moderately palatable plants like the American prairie, the advance of humans was only possible because cows possessed a traditional kind of biotechnology: they could turn green grass into red steaks. Modern America was created from the cow's stomach.

Farmers and animal breeders have worked hard on the cow's characteristics through a process that Charles Darwin called selective breeding: animals that produced large amounts of milk and good meat, were easy to tend, and were resistant to illnesses, were favored for reproduction, This is a creative achievement with high biocultural potential. The tight relationship between man and cow has created highly diverse landscapes.

353. Freeman Dyson, "Our Biotech Future, "*New York Review of Books,* (July 19, 2007): http://www.nybooks.com/articles/archives/2007/jul/19/our-biotech-future/.
354. Andrew Curry, "The milk revolution," *Nature*, vol. 500, (August 1, 2013): 20–22.

In the past decades, however, industrialization has distorted the relationship between cows and people in an unhealthy way. Modern-day cows have been turned into industrial meat production machines. They are no longer treated as animals, but as a mere step before packaged meat. Because demand for meat is so great, there are not enough pastures to feed grass to the entire cattle population. So, the majority of these animals spend their lives in factory outbuildings where they are fed corn or soy. Their body and muscle growth is like an internal assembly line. Biology is used here to drill the animals for their industrial existence: they are fed growth hormones and genetically modified plants and they are subject to genetic manipulation themselves to avoid diseases brought about by industrialization.

This has been taken to a new level since the complete genome of the cow has been decoded. It is now possible to discover which genetic factors affect the quality of meat and milk. Pre-implantation genetic diagnosis is used extensively to select embryos based on genetic criteria. Only the genes of embryonic cows that promise fast growth and juicy meat find their way from the test tube into the mother's womb.

Scientists have even broken through the biological barrier between humans and cows. Human genes in the cow's genome can turn the udder into a pharmaceutical bioreactor, a source of medical active ingredients. The production of antibodies, coagulation factors for hemophiliacs and even breast milk are just some of the goals in these experiments.[355, 356]

The pharmaceutical industry's wish list is long. Biomedical experts are already using cows as spare-part repositories for humans. Cardiac valves and glands from bulls are now being used in hospitals. The cow also serves as a model organism for new biomedical processes: for example, the US Department of Defense has developed the "Immunocow," a high-tech animal with a human immune system which is used to test new vaccinations against bioterrorists' weapons such as anthrax or smallpox, as well as new drugs.

355. James Robl, "Cloned transchromosomic calves producing human immuno-globulin," *Nature Biotechnology*, vol. 20, (August 12, 2002): 889–894 and Sanford Research http://www.sanfordresearch.org/researchcenters/appliedbiosciences/technology/tcbovine-technology/.

356. B. Yang et al., "Characterization of Bioactive Recombinant Human Lysozyme Expressed in Milk of Cloned Transgenic Cattle," *PLoS ONE*, vol. 6, no.3, (2011): http://phys.org/news/2011-04-genetically-cows-human-breast.html.

Many cows have significantly more complex, high-tech insides than the dairy factories where their milk is packaged or the slaughterhouses where they are killed.

Animal breeding has therefore entered a new phase. Cows are being altered on a massive scale and are even being humanized on a molecular level. However, this leads to worse rather than better treatment: cloned, genetically modified cows make people recoil rather than empathize. One of the researchers who develops genetic constructs for these animals had to admit that he has never actually seen or touched even one of the animals involved.[357]

The process of turning cows into technology takes place in anonymous, remote research institutes and factory farms, resulting in a downward spiral in our respect for these animals. Once they are turned into mass, the next step is to turn them into biomass. This could be just the beginning of a dangerous process in which our view of life is industrialized.

While our long-term companions—the dog and the cat—have a place in our hearts, we have a difficult, ambivalent and dishonest relationship with these latest man-made creatures. Just as people discharged chlorofluorocarbons and carbon dioxide into the atmosphere for years without considering the consequences, we are now collectively intervening in evolution without thinking about the scale of our actions. This is dangerous.

While the innovative aspects of genetic engineering are only just beginning to come to the surface of our collective consciousness, they are already being overtaken. A further, even more powerful step is about to happen: the most forceful example of how drastically and quickly humans can slip into a new role as creator (before they have really mastered it) is exemplified by the work being done at a private laboratory in America.

Less than 200 years ago, the first human experiment in synthesizing a substance present in living organisms achieved success. It was the year 1828 and Friedrich Wöhler, a Berlin scientist, demonstrated that it was possible not only to extract urea from the urine of an animal, but also to manufacture it using simple chemicals. In doing so, the scientist refuted a belief widely held in his time that a vital or magical force was necessary to

357. Ariel Schwartz, "Here's What Happens When You Put A Human Immune System In A Cow," *Co.Exist*, (September 25, 2013): http://www.fastcoexist.com/3017884 /heres-what-happens-when-you-put-a-human-immune-system-in-a-cow.

create a living organism. Urea consists of just seven atoms but because it is an organic substance, and therefore considered a "living thing," it seemed impossible to many of Wöhler's contemporaries, sacrilegious even, to attempt its synthesis. Many reacted to his experiment with disbelief, feeling that their religious views had been violated. But Wöhler had laid the groundwork for organic chemistry. Since then, biochemical research has been applied to every organism—protein by protein, gene by gene, and hormone by hormone.

Wöhler applied reductionism, one of science's most powerful epistemological principles, to the human body: dissection into its separate parts to get to the root of it. Wöhler would have marveled at what has happened in the early twenty-first century.

Man's first act of creation took place in a dish as round as the earth seen from the depths of outer space, and as bright blue as the planet's oceans— only much smaller. A petri dish and an incubator are all you need to create a culture of *Mycoplasma genitalium*. In May 2010, American biologist Craig Venter announced that he had created the first synthetic cell in living history in just such a petri dish. He spoke of "creation" and christened his life form, "Synthia."[358]

Synthia symbolizes more powerfully than ever that humans are starting to take on the role we used to reserve in our imaginations for a God, or gods. Stewart Brand, the founder of the *Whole Earth Catalog*, which Steve Jobs described as the forerunner of search engines, sees it as the chief task of our species to perform this role. In the 1970s, he made the casual remark: "We are like gods and we need to get good at it." Forty years later, his tone is a little more urgent: "We are like gods and HAVE to get good at it."[359] James Watson, the co-discoverer of the spatial structure of DNA, put it rather more flippantly: "If we don't play God, who will?"

Craig Venter chose mycoplasma as a model organism because it has the smallest genome of all living organisms. Mycoplasmas offer a small, manageable genetic landscape. At his institute in Rockville, Maryland, Venter and his colleagues synthesized the genome of this bacterium, molecule by molecule, and then implanted it into an existing cell envelope, resulting

358. Craig Venter et al., "Creation of a Bacterial Cell Controlled by a Chemically Synthesized Genome," *Science*, vol. 329, no. 5987 (July 2, 2010): 52–56.
359. Stewart Brand, *Whole Earth Discipline*, Penguin, 2010.

in a living organism. While Wöhler had begun with seven atoms, Venter's work involved millions of atoms. This bacterium is the site of the first genetic and molecular landscape created by a human designer.

Craig Venter has often played the pioneer in his life. In 1995 he was the first person to determine the genomic sequence of an autonomous living organism.[360] In 2000 he pressed ahead to publish a rough draft of the sequence of the human genome, putting on the defensive a worldwide consortium of hundreds of researchers who were attempting to do the same.[361]

In 2007 he mapped the first complete genome of an individual human, consisting of 2.8 billion base pairs—his own.[362]

One of Venter's favorite songs is by Crosby, Stills and Nash: "*Spirits are using me, larger voices callin' me. What heaven brought you and me cannot be forgotten.*" Venter, the freethinker and freebooter of genetic engineering, croons this tune as he sails across the oceans gathering bacteria from the depths to add to his genetic collection. Voices have indeed called him, saying: Transform something worthless—some bacteria that live in the ureter—into a microbial savior. And in so doing, strip the eco-freaks of their arguments against genetic engineering and the devout of their myths about god's role in creation.

First, the Synthia team practiced on a virus which is not considered to be really alive by biologists.[363] Only then did they dare move on to a proper living organism. In 2008, scientists at the Venter Institute synthesized the complete genome of mycoplasma in a yeast cell.[364] Two years later they created a living cell with the aid of this man-made genome.

---

360. Craig Venter et al., "Whole-Genome Random Sequencing and Assembly of Haemophilus influenzae Rd," *Science*, vol. 269, no. 5223 (28 July, 1995): 496–512.

361. Craig Venter et al., "Generating a synthetic genome by whole genome assembly: ØX174 bacteriophage from synthetic oligonucleotide," Proceedings of the National Academy of Sciences, vol. 100, no. 26, (23 December, 2003): 15 440¬–15 445.

362. Samuel Levy et al., "The Diploid Genome Sequence of an Individual Human," *PLoS Biology*, vol. 5, no. 10, (4 September, 2007).

363. Craig Venter et al., "Generating a synthetic genome by whole genome assembly: ØX174 bacteriophage from synthetic oligonucleotide," *Proceedings of the National Academy of Sciences*, vol. 100, no. 26, (23 December, 2003): 15 440–15 445.

364. Daniel Gibson et al., "Complete chemical synthesis, assembly, and cloning of a Mycoplasma genitalium genom," *Science*, vol. 319, no. 5867, (29 February 2008): 1215–1220 and Daniel Gibson et al., "One-step assembly in yeast of 25 overlapping DNA fragments to form a complete synthetic Mycoplasma genitalium genome," *Proceedings of the National Academy of Sciences*, vol. 105, no. 51, (23 December, 2008): 20 404–20 409.

This represents a quiet but nevertheless deep rupture with billions of years of evolution: it was not the result of blind forces of nature—Darwin and Mayr's much-acclaimed chance—but came from the single-mindedness of one human being. Venter's goal, and that of many other representatives of synthetic biology, is to create a completely man-made life form. What is still required is to synthesize cell walls and cellular fluid, which will take some time—but probably not long enough for us to rest assured. When Venter first unveiled Synthia to the public, he had this to say about the creation process from thought to computer model to laboratory: "This is the first self-replicating species that we've had on the planet whose parent is a computer."[365] But that was false modesty. Even if it is unclear whether Synthia really is a new species in formal terms, at least one of its parents is human.

In nature, new genomes are constantly being created from separate components; it's a process that takes place a trillion times a day. But imitating this process in a laboratory puts it under human control. Rationality and planning come into play but so do darker traits, like megalomaniac obsession.

If Venter retired today, he would have already achieved considerable fame. He has produced the most complex living structure ever to be consciously designed. But his aims are even higher. He intends to program his synthetic bacterium so that it can sustain itself with carbon dioxide, a greenhouse gas, and create energy-rich hydrogen. Venter promises that this could solve the human energy problem in a biosynthetic way. And with this colossal, microscopic step towards environmental recovery, he wants to ensure that he is truly unforgettable.

Whether he succeeds depends on how well the other cell components besides DNA can be biochemically manufactured and combined. The side effects are unknown: what happens if the bacteria devour carbon dioxide too efficiently? Or if they escape into the open and manage to survive? Then Venter might be immortalized as the man who introduced the new Ice Age instead of putting a halt to global warming. No matter what happens, he will go down in the annals of history as a great scientist: Synthia ranks among the most significant chemical experiments ever carried out.

365. http://www.ted.com/talks/craig_venter_unveils_synthetic_life.html_.

Venter sees himself as the obedient servant of Wöhler's program. If he succeeds in chemically synthesizing an entire organism, it will be proven beyond a doubt that the transition from dead matter to living being does not require divine intervention.

In the twentieth century, researchers were still restricted by the need to carry out elaborate manual work to isolate single proteins or genes from specimens in order to be able to examine them. Nowadays, robots can read the virtually complete genome of a life form in a short time. It cost several billion dollars and took from 1990 to 2000 to sequence the first complete genetic blueprint of a human; nowadays it costs a few thousand dollars and takes a matter of days.[366]

The events taking place in the Venter Institute are like a continuation of the opening sequence in Stanley Kubrick's world famous film, *2001: A Space Odyssey* in which a bone that is flung through the air by primeval humans as a weapon is transformed into a spaceship, and then into a pen by a series of skillful film cuts. In the Anthropocene, the cudgel with which early humans in the Pleistocene hastened the extinction of entire hordes of animals has transformed into a pipette with which scientists trickle an elixir of creation into a petri dish and generate life.

Craig Venter is a pioneer, and just one member of a broad movement among scientists who want to conquer life. While Venter and other genetic engineers are acting on molecular levels, stem cell researchers are trying to create new forms of life in a different way. The discipline of stem cell research picks up where *2001: A Space Odyssey* left off. In the final scene of the film, the astronaut floats as a fetus through space, rejuvenated by an unknown force.

Similarly, laboratory scientists have managed to restore mature adult body cells to an embryonic state. This means that in the future, humans will be able to turn back the biological clock. Cells will regain the strength that they had shortly after their individual lives began, shortly after the egg and sperm of their mother and father fused. This is the strength that creates an entire organism with all its various specialized tissues.

366. Wetterstrand KA., "DNA Sequencing Costs: Data from the NHGRI Genome Sequencing Program (GSP)," available at: www.genome.gov/sequencingcosts. Accessed December 19, 2013.

For a long time, there was a prevailing dogma that eggs and sperm, as opposed to neurons or liver cells, were fundamentally different, and that scientists would never be able to cultivate the gametes of life. This dogma was crushed in 2003 in the laboratory of the German stem cell researcher, Hans Schöler.[367]

The discovery happened by chance. An experienced technician saw a strange, round formation swimming at the edge of the stem cell culture in front of her. They turned out to be egg cells. A chemical that was mixed with the cells as a growth factor had catalyzed an unforeseen effect. Since this incident, the germ cells of life have been available in synthetic form in private and state-run laboratories, where curiosity and greed, reason and imperiousness, genius and human delusion all build to a crescendo.

This unexpected discovery kicked open the door to the control room of life that had remained closed to humanity until that moment. Inside, there are "levers" that can produce any number of nerve cells, liver cells, skin cells, or any other cell type. There is also a "steering wheel" that permits biological time to be turned backwards to embryonic stages. Genes that have a decisive influence on the form and function of a living organism can be turned on or off.

Now scientists are beginning to practice with these instruments. For the moment, everything still takes place in the laboratory; scientists are still fumbling with the levers, and results have not yet lived up to expectations. But slowly and surely, they are making progress, and their dexterity at the controls is becoming more proficient. The general public is also beginning to realize the life forces that are being released. The scene where a man-made homunculus calls his maker "Father" in *Faust*, Goethe's literary classic from 1808, is now becoming a reality. And while Woody Allen's wish, "I don't want to achieve mortality through my work, I want to achieve it by not dying," might not be feasible quite yet, biology gives the Anthropocene a further deep dimension. What is being created in today's laboratories could change the course of biological world history and rewrite the rules defined by Charles Darwin, Ernst Mayr and other evolutionary biologists.

367. Karin Hübner, Hans Schöler et al., "Derivation of oocytes from mouse embryonic stem cells," *Science*, vol. 300, no. 5623 (23 May 2003): 1251–1256.

The human mind is becoming a powerful force alongside blind chance in the evolutionary process and it is beginning to influence that process. Methodological progress is so great that the general public should prepare itself for some radical scenarios: sooner or later, 4D models of organisms will be created together with the blueprint of their biochemical synthesis.

We are still in the realm of possibility at this stage but phenomenal progress is being made, and fast. That's why it is important to check the kind of minds and intentions at work. Are biologists really engineers, as is expressed in the word "genetic engineering?" Is it their task to incorporate life into the logic of industrialization? Or are they stewards of life, who go about their quest to understand biology with empathy and respect? These questions will be strong shaping forces in the Anthropocene. Without consensus on the aims and limits of cultivating synthetic life, this field will soon simply reflect the ambitions of individual scientists or the profit-hunting instinct of large companies. Therefore, it is imperative to discuss how state and private laboratories can become democratic forums so that the creatures they generate in the future come from a mature consciousness rather than being jolted to life and shocking society.

The danger that artificial life could escape from laboratories in the process of synthetic biology is just one aspect of the problem. The greater danger is that scientists change fundamental attitudes to life for the worse: an industrial biology that produces animals as if they were flat-screen TVs, or is prepared to risk any kind of cruelty in genetic engineering for medical discovery and treats living creatures as mere "technology platforms," embodies a brutish attitude towards life. While this attitude might appear to be sophisticated and academic at first, it could soon blaze a trail through society that is detrimental.

Private companies, which Venter's Creation project ultimately represents, have often been the incubators of important, positive innovations. Companies and individual pioneers generate not just the capital they need for their research, but they often work under some of the best intellectual conditions for bringing innovation into the world.

However, these innovations enter the world in a very specific way—as products or commercial projects. This is at odds with the fact that what happens at the Venter Institute and Synthetic Genomics Inc. in La Jolla,

California, represents one of the most sensitive research projects of all time, and the entry into a new dimension of life and human power.

The project was financed by several hundred million dollars' worth of investment from oil companies. Is this some kind of catharsis? Are the mass producers of oil and greenhouse gases starting to shepherd bacteria in order to milk clean energy? True, a conversion on this scale is what is desperately needed. But perhaps these companies should first deploy existing renewable energy technologies on a grand scale and not put their bets on a hypothetical breakthrough technology that involves fiddling with the fundamental forces of life.

If mankind wants to create synthetic life, the first creature resulting from this should not be an industrial product. It should drift like a tender nymph from the brains of scientists and be an inspirational, artisanal marvel rather than a kind of replacement drug for our addiction to cheap energy. As it stands, the Synthia project holds up a mirror to our society: artificial life is being created to fill our gas tanks rather than for something noble such as the creation of a settlement on one of Saturn's moons.

An even harder question in this respect and one that affects people directly is: *who* is in control? How should a society that uses nature as mere biomass and raw material deal responsibly with the knowledge of how to synthesize life? Are modern democracies strong enough to use the powers of creation moderately and in a meaningful way? The eugenic crimes of the twentieth century are a strong warning of just how wrong developments can go. Luckily, scientists like Venter are not under the least suspicion of thinking along such lines. He himself has called for an intensive ethical debate on his research in order to recognize and sound out misguided developments in time.

When humans take their own evolution in hand, where should the borders lie? The announcement that illnesses such as Huntington's disease (a hereditary disorder that affects few families and leads in most cases to a slow, agonizing death in mid-life) will be wiped out in the future through a combination of artificial insemination and genetic testing, allowing healthy babies to be born to these families, sounds positive and life-affirming. But where does intervention stop? Will research into behavioral genes be used in molecular selective breeding? Will humans in the

Anthropocene no longer be products of nature at their biological core—products not of chance but products that go on to create themselves?

The progress of modern bio- and genetic technology make it seem likely that the sixth day of Genesis will turn into a science fiction plot that runs backwards: at the end, we release our very own Adams and Eves. Christianity's claim that man was made in God's likeness would manifest itself in an unexpected way. But the day that humans recreate their own image from individual molecules in the same way that Craig Venter has created Synthia, is still some time away. Until then we have the chance to find the right conditions for this knowledge and decide in whose hands power over life will end up. Will those be the hands of individual companies, like the Tyrell Corporation in Ridley Scott's film *Blade Runner,* or research strategists at the defense agencies of the American or Chinese? Or in the institutions of a democratic public? Will we be in a position to handle the synthetic creatures of the future with empathy and respect?

So far, the response to Venter's appeal for an ethical debate has been subdued. Again, intellectual and emotional understanding lags behind technological advances. In order to explore and understand the new nature of the Anthropocene, another Charles Darwin or Ernst Mayr is needed.

During our conversation at the old people's home at Badger Terrace, Mayr defended the traditional evolutionary theory with wise words: "I respect people that say nature cannot all be down to chance," he said. "But in the light of my own knowledge, I have to say that all this was purely chance, even if I don't like the fact."[368] Well, this analysis is certainly true for the past billion years, but the future of evolution will be different. Darwin and his apostle, as Mayr was called, formed their ideas during expeditions in world regions that were still primarily wild at the time. Biologists today have an entirely different world before them: a world in which industrialization first steamrollered the living environment, and next, the concept of life itself. There is not even an outline of which ideas a new Darwin could sketch concerning the directed evolution of the future. Darwin showed that a creator was not necessary for biodiversity. The next Darwin will have to make sense of a world in which nature is transformed into

368. Joachim Müller-Jung and Christian Schwägerl, "Darwins Apostel," op. cit., see Footnote 344.

culture and people test out the role of creator. He will be an explorer of the Anthropocene.

Charles Darwin was inspired by the flower meadows that his botanist uncle Erasmus introduced to him. Ernst Mayr was inspired by his observation of birds. Perhaps future biologists will be impressed by the barren wastelands of a Chinese settlement on the fringes of a city, or by the sight of crow's nest made from clothes hangers, or a Texan fracking landscape. They will encounter life forms like mycoplasma that have been created by the human mind rather than a natural struggle for survival. They will have to develop views on highly fundamental questions: should life continue to be industrialized? Or should it be the other way round—should industry go through a process of biologization, and production platforms be transformed into something more attuned to life? Will people in the future ascend into brutal "Supermen" with the help of genetic technology and brain research, or will they use their new knowledge to strengthen their empathy with all creatures?

Researchers of the future will be able to draw on enormous databanks in which gene sequences, protein activities and the biological scripts of thousands of species and millions of individuals are stored. They will understand the brain in a depth that seems impossible to those alive today. The fact that it is possible for us to make live observations of the individual nerve cells of laboratory animals gives us an inkling of what might be possible in the future.[369, 370]

If anything demands farsightedness and a sense of responsibility, then it is surely our conscious design of life forms that, if released, might outlive *Homo sapiens*. The dark potential for evolutionary biological accidents in these "life factories" is as great as the danger that life will be defined for a purpose or that the heresy of Social Darwinism will resurface in a new form.

What can science do to prevent an essentially negative development? It is less a question of *what* scientists do than *how* they go about and discuss

369. Karl Deisseroth et al., "Multimodal fast optical interrogation of neural circuitry," *Nature*, vol. 446, (5 April, 2007): 633–639.

370. Georg Nagel et al., "Fast manipulation of cellular cAMP level by light in vivo," *Nature Methods*, vol. 4, no. 1, (January 2007): 39–42.

the issue. No one should or can put the brakes on human curiosity. Therefore, the culture in which this curiosity unfolds is all the more important.

How might a different approach look? There is a huge difference, for example, whether millions of genetically modified mice are regarded as mass products and are disposed of as organic waste after use, or whether new forms of respect and gratitude develop. This could take the form of a yearly rite in the biomedical community, for example, in which these animals receive special feed and are praised in speeches, a kind of "Thank the Mouse" day. This will change very little for the mouse immediately but it might change the development of short and mid-term research.

The same can be said of cows. It isn't necessary to go to the extreme of declaring the cow sacred, as it is in India. But these animals, for which we have so much to thank, should no longer be hidden, made taboo or even despised because we have genetically modified them. School classes should regularly visit laboratories and animal factories to experience these animals close up and develop empathy for them. Animal adoption schemes could be set up, whereby classes of children or families accompany the lives of animals, and therefore ask themselves the price that society is prepared to pay for the mass consumption of beef and the blessing of biotechnology. As it stands, parents feed their children milk and beef day in and day out without so much as having milked a cow or seen it being slaughtered.

It also makes a difference whether scientists constantly emphasize that people and animals consist "only" of matter—or whether an intrinsic value is attributed to this matter. The word "only" echoes in all the important present-day scientific debates, such as the defense of free will. With noticeable *schadenfreude*, for example, some neurobiologists cite as evidence the fact that human decisions and consciousness are "only" based on mechanistically explained neuron circuits; their opponents shudder at this devaluation of humanity: "Is God just a storm of neurons in the end?" they ask provocatively. The "only" fails to recognize what a fantastic storm it must have been to bring forth such grand ideas, and the incredibly complex processes that take place in even the simplest of life forms, processes that leave any human endeavor looking primitive. Getting beyond the "only" is one of science's most important tasks, in my opinion. For, in the end, how will the natural sciences demand respect and empathy for something that they describe and treat disparagingly?

Disdain for material life is deeply rooted in the body-mind dualism of Western philosophy, in the clear separation between the (base) body and (elevated) soul. It was encapsulated by the seventeenth century philosopher and scientist René Descartes in his work *Meditations on First Philosophy*. The brilliant French thinker not only coined the phrase, "I think, therefore I am," but also put forward the notion, rooted in science, that only humans have an immaterial soul that connects them to God. Animals and the human body, on the other hand, were merely unconscious *automatons* according to Descartes. "A variety of dualisms were proposed to account for many of the same complexities of the world," writes the neurobiologist Matthew D. Liebermann, "and yet none really stuck until Descartes' version."[371]

Descartes accomplished great things for nascent rationalism, and helped promote skepticism as a scientific principle. This earned him the attention of the Inquisition and his writings were banned by the Vatican; despite which he managed to achieve lasting fame. However, the image of the body that Descartes thought up in the seventeenth century has had catastrophic consequences for respect for life. He contributed to today's attitude that scientists consider organisms to be machines. The machine metaphor has become dominant in describing ourselves. This has contributed to bizarre forms of alienation. For example, brain researchers systematically talk as if unconscious molecular processes in the brain were somehow disproving the existence of free will and that therefore we are zombie-like creatures. Why don't they examine how unconscious molecular processes *enable* the freedoms we enjoy? A similarly misleading deterministic message is that humans are mere puppets manipulated by their genes.[372] I wonder why people like Richard Dawkins don't talk about the dense social fabric that links genes, proteins and the environment.

When the first complete version of a human genome was announced in 2000, scientists like Venter spoke of "the book of life" being opened. But

---

371. Michael D. Lieberman, "What makes big ideas sticky?" from: *What's next? Dispatches on the Future of Science*, (New York, Vintage Books, 2009) and M.D. Lieberman, "Social Cognitive Neuroscience: A review of Core Processes," *Annual Review of Psychology*, vol. 58 (2007): 259–289.

372. Richard Dawkins, *The Selfish Gene*, Oxford University Press, 1976.

since then, they have come a long way and have realized that life is considerably more complex than a sequence of chemical building blocks. It is not so much a book as an adventure along the lines of *Alice in Wonderland* in which new doors and dimensions keep opening up. The secret of humanity was not disclosed along with the human genome sequence. Not even the secret of a fruit fly's existence is contained in the sequence of its four genetic components. The genome is not a mechanical engine that simply has to rattle into action in order for life to move along. Nor is it a type of software booted up by an organic robot. The notion that living beings simply perform what their genes have long determined is now considered far too simplistic, "naïve," as Venter says these days.

The scientific method has been superb at producing billions of snapshots of what's going on inside of organisms. The difficulty now is to rearrange the puzzle, to put the pieces back together to form a coherent "movie of life." That's why after decades of breaking up life into ever-smaller bits, scientists are now taking a renewed interest in entire cells, entire organs, and entire organisms. DNA reductionism—the idea that life is controlled just by the sequence of base pairs—is on its way out. A much richer picture emerges, one that links DNA, proteins and the environment: Genetic material produces messenger molecules that yield proteins; these in turn react to environmental stimulants that guide the activity of genes and produce new messenger molecules. Among the 1,712 genes in beer yeast alone, scientists have already determined 5.4 million paired reciprocal effects.[373] With the help of substances that show up on MRI scans, chemical reactions on the inside of the body can be observed in real time from the outside. Gene chips make it possible to "see" which genetic material is active and which lies dormant. New methods such as optogenetics, whereby genes are turned on and off using light impulses, are paving the way to a script on living processes.

These methods reveal whole worlds beyond DNA determinism: Epigeneticists now study how environmental stimuli influence when which gene is active via the molecule methyl. Entire complexes of protein react to psychological and physical stimuli from the brain and environment,

---

373. Michael Costanzo et al., "The Genetic Landscape of a Cell, Science," vol. 327, no. 3964 (22 January 2010): 425–431.

thereby acting as intermediaries between the outside world and genetic code.[374] Some changes that find their way from the psyche and environment into the organism are even hereditary.[375] This means that no clear separation can be made on a genetic level between body and mind, environment and genes, conscious and unconscious, as Descartes' simple dualism claimed. After decades during which evolutionary psychologists have placed the focus on the selfishness of the individual, biologists are realizing more and more that the sum of a person is wired to sociability rather than selfishness.[376]

It is not the job of scientists to come up with a positive, uplifting and illuminating image of humans and nature just to comfort people. But likewise, it is not their job to block out the positive, uplifting and illuminating knowledge of humans because it doesn't tally with the old tenets of determinism and the machine metaphor.

Descartes' impulse had great impact on modern science and consequently influenced the early Anthropocene. He, of all people, drove a solid wedge into what used to be universal science, and divided the natural sciences from the humanities. For centuries, this segregated the machine builders from the philosophers, the botanists from the psychologists, the geneticists from the ethicists. Modern science still lives in a Cartesian mindset with divided spheres for the inferior, exploitable material world on the one hand, and the superior intellectual world on the other, upon which human dignity is founded. There is no provision for the dignity of nature. For this reason, the Enlightenment that followed Descartes did not lead to a cooperative treatment of nature but instead unfurled itself as a program of subjugation.

In the nineteenth century, Alexander von Humboldt emerged as an opponent of Descartes. The goal that he expressed for the rulers of his time was not the one-sided wealth of Europeans through their brutal perse-

374. Rudolf Jaenisch and Adrian Bird, "Epigenetic regulation of gene expression: how the genome integrates intrinsic and environmental signals," *Nature Genetics*, vol. 33, (2003): 245–254.

375. See Eva Jablonka und Marion Lamb, *Evolution in four dimensions: genetic, epigenetic, behavioral, and symbolic variation in the history of life*, Cambridge, MA: MIT Press, 2005.

376. For a summary in layman's terms, see also Yochai Benkler, *The Penguin and the Leviathan, How Cooperation triumphs over self-interest*, New York: Random House, 2011.

cution as colonial powers, but equal wealth across the globe. He did not plot the subjugation of nature but the use of its principles. In the midst of the colonial era, and the wave of industrialization that began to emanate from England during that period, Humboldt stood for the notion of a different, more humane and more ecological modernity.[377] The German scholar and naturalist regarded it as science's aim to develop a sustainable philosophy of nature, a doctrine of *Weltorganismus* or "world organism" that deserved respect. His life work amounted to "reconciling" man and nature.[378] He never regarded animals and people as if they were mere machines. In the *Weltorganismus*, Humboldt considered all living creatures as being connected.[379]

The early Anthropocene would have been very different if Europe and America had assimilated this inclination more thoroughly, especially since Humboldt was in close contact with the rich and famous of the time, and even maintained a close friendship with such luminaries as the American President, Thomas Jefferson.[380]

Humboldt's holistic way of thinking is available to us today as a path through the Anthropocene. When we pass each other in the "buyosphere" of the twenty-first century's supermarkets, when the creations from selective breeding dominate our landscapes and new life forms escape from biologists' laboratories, our appreciation of life will be challenged. Will we subjugate it to purely utilitarian thinking or enhance it in other ways, thus developing it for a world beyond ours in which nature has become culture?[381]

People, as opposed to animals, are granted dignity in Western culture—an inviolable, transcendent protection in our dealings with one another and through the social institutions of living together. To sepa-

377. See also Ottmar Ette, *Weltbewußtsein. Alexander von Humboldt und das unvollendete Projekt einer anderen Moderne*, Weilerswist, Weilerswist-Metternich: Velbrück Wissenschaft, 2002.

378. Alexander von Humboldt, *Views of Nature: Or Contemplations on the Sublime Phenomena of Creation*, Cambridge University Press, 2011.

379. Alexander von Humboldt, *Cosmos: A Sketch of the Physical Description of the Universe*, Kindle Edition.

380. Laura Dassow Walls, *The Passage to Cosmos: Alexander von Humboldt and the Shaping of America*, University of Chicago Press, 2009.

381. The term "buyosphere" was coined by Daniel Goleman in *Ecological Intelligence*, New York: Broadway Books, 2009.

rate humans from animals in this way was perhaps a necessary cultural operation to counter our tendency to think that "Man is a wolf to his fellow man," in the words of the ancient Roman playwright Plautus. But now that the domestication of the wolf has led to something completely different—the dawning world of the Anthropocene and directed evolution—it's time for the next stage of cultural evolution: one where we do not regard our species as the ruler of earth but as part of a vibrant *Weltorganismus* that deserves its own kind of dignity. Scientists like the consciousness researcher Giulio Tononi, already ask for a whole new perception of nature, in which all living beings are connected by a diversity of conscious states. The "organic-unfathomable" is what Ernst Mayr called nature during our visit. He regarded it with reverence his whole life long.

TEN Earth Economy

THE CENTRAL AFRICAN REPUBLIC, with an annual gross domestic product of 900 US dollars per capita, is one of the poorest countries in the world—and this is extremely obvious in the capital of Bangui. Many people show signs of not having had enough to eat; their cheeks are sunken and they walk slowly, clinging to the walls. It happens that people suffering from AIDS lie down in the gutters to die. Numerous buildings have been destroyed by past civil wars or simply crumbled through neglect, among them ministries, law courts and prisons. Instead of going to school, children go looking for food or else beg from the foreigners that have ended up here, mostly Russians who take diamonds out of the country, or aid workers like the man with whom I was traveling on this trip.

The poverty here is absolutely appalling, and so the reason for my visit to this country seemed absurd at first: I had come to see the wealth of Central Africa, its natural wealth. The Congo Basin has the second largest tropical rainforest in the world, and in comparison with South America and Asia, the rate of deforestation is relatively low. In these rainforests live enormous concentrations of rare, large mammals.

A few days after leaving Bangui, we arrived in the Dzanga-Sangha region, in the southwest of the country. A team from the World Wildlife Fund (WWF) and a forest guide from the BaAka Pygmy people showed us the way to a saline—a salty, open clearing in the forest—where we saw dozens of elephants taking their evening mud baths. Not much further on, we encountered a very relaxed extended family of lowland gorillas and our guide showed us the tracks of forest buffalo, giant forest hogs and bongo, a variety of antelope.[382]

382. For further information on the Congo Basin and protected areas there, see www.dzanga-sangha.org and www.theguardian.com.

Brutal human poverty on one hand, tremendous natural wealth on the other—the contrast could not have been greater between our time in the forest saline and the oppressive hours spent in Bangui. Did one have to do with the other? Had Central Africa failed to log its forests in time or sell its minerals quickly enough? The reasons why the Central African Republic is so poverty-stricken are highly complex: in part, they go back to the French colonial era, in part, they stem from the corruptibility, brutality and incompetence of the nation's ruling elite, and in part, they are due to the country being positioned far away from trade routes, among other contributing factors.

The aid worker with whom I was traveling was a former top official in the German government whose mission was to explore ways in which this poor patch of the world could earn money from its natural riches without scarring the country with deforested clearings or uranium mines. This seems almost impossible in present circumstances. Our current financial system ascribes as little economic value to a living rainforest in the Congo as to vast Russian bogs, Caribbean coral reefs abundant with species or a South Pacific marine area teeming with fish. Value is only generated from the "use of natural resources"—that is, only when trees are felled to make wooden floors for middle-class homes, when bogs are dug up and their peat used to fertilize blossoming balcony plants, when coral reefs are transformed into construction material for hotels, or when trawlers catch fish by drifting their vast nets through the ocean.

Value primarily arises in the financial capitals of the world: London, Hong Kong, Frankfurt, Tokyo, and above all New York, where a single business dinner with clients can easily cost 900 US dollars and where an investment banker or trader can earn as much or more money with a couple of computer clicks in under a minute than 4.5 million Africans can in an entire year. Value is generated through interest on money lending, through betting on the future creditworthiness of countries, or betting on whether other bets will come off.

By contrast, from an economic point of view, the basis of life on earth has no value, not even the precious saline in the Dzanga-Sangha forest.

Classical economics has built an impenetrably high wall around itself, a wall between the "inside" and the "outside," between ecology and economy. On the inside there is money, capital investment, goods and labor; on the outside there is nature, the ecosystem, waste, and dangerous changes to the environment. This artificial division allows us to build up virtual wealth that only works because its cost and damages are never shown on any balance sheets. Economists have a word for what holds the world together at its core: they call ecological functions such as oceans, forest, and atmosphere "externalities." But the problem is that, in the reality of the Anthropocene, there is no more "outside," and the externalities have long since become internalities. The rainforests, for example, make a decisive contribution to the regulation of global climate, which in turn has a significant effect on whether investments in buildings in New York will be swept away by the next superstorm. However, these interdependencies do not appear on financial statements. Classical economics is still blind to the fact that economy will become ecology in the Anthropocene because without the services of the oceans, forests, climate and water balance, a sustainable economy is not possible anywhere in the world.

The financial crisis that started with the collapse of the Lehman Brothers bank in 2008 demonstrated what happens when relations on the inside diverge too greatly from those on the outside. For years, banks had forgiven mortgage loans so they could rake in commissions. They earned money by hawking bundles of bad loans to unknowing investors. On paper, in other words on the inside, the banks' business looked good and until recently, enormous bonuses flowed into the pockets of bankers who sold bad loans. But outside in the real world, a catastrophe was brewing as millions of people were not in a position to repay their loans. The house of cards then collapsed.[383, 384]

A similar thing is happening on a much greater scale with the economy's treatment of the living world. Inside, on paper, there are enormous growth rates and impressive profits. But outside, a dwindling sum of eco-

383. For a deeper insight into the mechanisms of the financial crisis see Michael Lewis, *The Big Short—inside the doomsday machine*, New York: Norton, 2010.
384. Robert M. May et al., "Ecology for bankers", *Nature*, vol. 451, (February 21, 2008): 893–895.

logical capital is at our disposal in the form of food, drinking water, raw materials and climate, no matter whether we are poor African farmers or super-rich bankers in New York City.

By depleting fish stocks, reducing global forest areas, or overloading the atmosphere with greenhouse gases, we are taking out a bad loan from the ultimate bank, the Earth. We feign wealth and convince ourselves that we have a planet at our disposal that is much bigger than it really is—just as the borrowers in America and Europe thought they could afford over-sized houses. In the long run, this doesn't work.

To take out a healthy loan on the ocean would mean knowing its carrying capacity and establishing restrictive quotas for regional and global fishing that allow stocks to regenerate. To take out a healthy loan on the forest would mean clearing only so much forest as to allow for healthy and biodiverse regrowth.

To take out a healthy loan on the climate would mean to only emit as much carbon dioxide as the atmosphere and oceans can absorb without dangerous changes.[385]

The interest costs that we would have to pay on healthy loans from the earth's capital would manifest in the form of restraint, consideration, innovation and good management. It would include political obligations for corporations and citizens that ensure that the oceans, the forests and the atmosphere can preserve their complex cycles. It would include protection zones and periods in which natural capital can regenerate or stabilize in order to become available once again as loan capital. Humanity has been helping itself to natural capital on a grand scale for decades, without paying interest or amortizing costs. Bankrupt banks and countries can recover but in the case of bankrupt ecosystems, it isn't that simple. It will

385. On the economic aspects of marine protection, see Boris Worm, "Impacts of Biodiversity Loss on Ocean Ecosystem Services," *Science*, vol. 314, no. 5800, (November 3, 2006): 787–790; Steven Murawski et al., "Biodiversity Loss in the Ocean: How Bad Is It?," *Science*, vol. 316, no. 5829, (June 1, 2007):1281; Walter Garstang, "The Impoverishment of the Sea. A Critical Summary of the Experimental and Statistical Evidence bearing upon the Alleged Depletion of the Trawling Grounds," *Journal of the Marine Biological Association of the United Kingdom*, vol. 6, (July 1900): 1–69, 1900; Daniel Pauly et al., "Fuel price increase, subsidies, overcapacity, and resource sustainability," *ICES Journal of Marine Science*, vol. 65, no. 6, (April 2008): 832–840 and Jeremy Jackson, "Ecological extinction and evolution in the brave new ocean," *Proceedings of the National Academy of Sciences*, vol. 105, suppl. 1, (August 12, 2008):11458–11465.

take centuries, perhaps millennia or even longer until their complexity or former capacity is even partially restored.

But ongoing financial crises have not triggered more profound reflection. When the bubble burst, the protective walls of banks held up amazingly well due to the industry's good relations and embedded staff at the centers of political and military power. The economic follow-up costs of multiple wrong decisions were not borne by the perpetrators but by the middle class and the poor, in the form of unemployment, tax-funded rescue packages, as well as reductions in state investments for education and research. The banks even profited from enormous sums of taxpayers' money in order to make the walls around themselves higher.[386]

While 2.5 billion people worldwide have to live without a toilet, 1.6 billion have to live without electricity[387] and those in countries like the Central African Republic, cannot escape political chaos and poverty, the 400 richest Americans' fortunes grew in one year by an average of 800 million to 5 billion dollars at the end of 2013, making a total of 2 trillion dollars.[388]

What could be a better way to do business with the deepest sources of global wealth than to give the poorest of the poor the foundation for independent economic development?

The development strategist with whom I traveled to the Central African Republic wanted to find an answer using a concrete example. We visited an eco-lodge run by the WWF built on the concept that rich nature lovers come to the region and pay good money to see forest elephants and gorillas. A large proportion of the proceeds are then used to benefit the local people in the forest region by paying for schools and medical treatment. This model has been practiced in many areas of the world with more or less success in individual cases. But this method falls too far short

386. Most of the main agents of the financial crisis, such as the investment bank Goldman Sachs, are bringing in record profits once again and experts are already giving warnings of a new credit bubble.

387. UNDP, "The Energy Access Situation in Developing Countries: A Review Focusing on the Least Developed Countries and Sub-Saharan Africa," (Paris, 2009) and UN News Center, "Deputy UN chief calls for urgent action to tackle global sanitation crisis," (March 21, 2013): http://www.un.org/apps/news/story.asp?NewsID=44452&Cr=sanitation&Cr1=# .UU_G_BySV3-.

388. Luisa Kroll and Kerry A. Dolan, "The Forbes 400—the richest people in America," (September 16, 2013): http://www.forbes.com/forbes-400/, retrieved December 21, 2013.

of the mark: the multitudes of rich travelers necessary to conserve the whole world's critical ecosystems do not exist; and if they did and all set off for their ecological dream destinations, their carbon footprint from additional air flights would be immense.

A much greater step is necessary to balance the earth's economy and to protect it from a Lehman-style crash: functioning ecosystems have to be ascribed their own economic value so that they appear on the positive side of company or government balance sheets. This has to happen while they are living and sustained, not when they have already been destroyed.[389] My traveling companion was therefore having discussions in the border regions of the Central African Republic, the Congo and Cameroon on how it might be possible to put to commercial use the Congo forests' positive influence on the world climate. He envisioned making a marketable service out of the fact that the Congo's forests bind a share of the carbon dioxide that industrial states blast into the atmosphere from their power stations and automobile emissions: a global air-conditioning service charge, so to speak. According to this model, rich industrial nations in the future would pay to keep using the Congo forests as a global water reservoir, carbon dioxide storage site and biodiversity hot spot. As climate researchers have calculated that a maximum of two tons of carbon dioxide emissions per year and per capita is permitted to accumulate over the twenty-first century, a price could be set for surplus tons (fourteen in the USA and eight in Germany as of 2012) which would be paid to those who cause fewer emissions or absorb greenhouse gases from the air with the aid of their ecosystems.[390, 391]

The idea of paying for "ecosystem services" hearkens back to the common roots of economy and ecology. These two words are defined nowadays as irreconcilable opposites but both of them derive from the same Greek root: *oikos*. This is no coincidence; on the contrary, it is highly meaningful and relevant.

389. Philip Bethge, Rafaela von Bredow and Christian Schwägerl, "The Price of Survival: What would it cost to save Nature?," *Spiegel Online*, (May 23, 2008): www.spiegel.de/international/world/a-554982.htm.

390. Ralf Antes et al., *Emissions Trading: Institutional Design, Decision Making and Corporate Strategies*, Berlin: Springer, 2011.

391. Chinese Academy of Sciences, *Carbon Equity—Perspective from the Chinese Academic Community*, Beijing, 2009.

Oikos was the Ancient Greek term for a household, a family community, and by extension, a hearth or oven. The oven is a place where raw nature is transformed into edible culture. Because the Ancient Greeks took intervention in nature very seriously, a culture of sacrificial offerings arose surrounding the oven to apologize for the extraction of natural wealth and restore order among the gods. From these sacrifices arose the first form of money, which was used to replace sacrificial animals.[392] So the word *oikos* reflects the way in which people perceived and managed their close relationship to nature from a very early stage.

Even more than 2,000 years after the shared roots of ecology and economy grew in Greece, the ties between them have not been cut. When the polymath Alexander von Humboldt was a young Berliner in the mid-eighteenth century, his mother insisted he study financial management. Alexander wanted to embark on a career in science or the army and no way did he want to become an economist. But to his own astonishment, his economic studies at the University of Frankfurt an der Oder began with botany, based on the economic theory of Johann Beckmann, an Enlightenment philosopher who coined the word "technology" and turned his attention to the protection of resources. Due to this experience, Humboldt became a great naturalist, but he always remained an economist too and began his career as a mining official.[393]

Imagine if the MBA students being churned out of universities these days by the dozens to work in investment banks had first to plan a herbarium or receive practical training at a National Park! An absurd idea? No, in fact it's just what is needed for the economists of tomorrow to understand natural capital and thus factor it into their project models. For a while now there has been a small but growing economic reform movement that espouses tearing down the artificial walls between economy and ecology. One of its doyens is Pavan Sukhdev who used to manage the Global Market Center of the Deutsche Bank, and oversaw its growth from 53 to 350 employees before he devoted himself to an entirely new field of business, the economy of nature. Sukhdev, who advises institutions like

392. Christina von Braun, *Der Preis des Geldes. Eine Kulturgeschichte*. Berlin: Aufbau-Verlag, 2012.

393. Alexander von Humboldt, *Aus meinem Leben*, Munich: C.H. Beck, 1989.

the European Commission and the United Nations Environmental Program (UNEP), says that we misunderstand the nature of value and have a false understanding of the value of nature. In classical economic textbooks, the value of living nature amounts to nothing unless turned into some sort of product. However, nature's value is not some kind of idealistic notion in new economic theory but something very real and tangible.[394, 395]

A central problem is that contemporary business studies courses strongly devalue the future in favor of the present with a so-called "discount rate." This establishes, by way of example, that a forest rich in species will only be worth a seventh of its present value in fifty years' time and thus favors rapid exploitation.[396, 397]

New types of economists like Sukhdev see a gigantic expropriation of future generation's resources taking place: "For example, when a sector like palm oil is destroying rainforests to grow palm oil, that means that they are losing the value of rainfall that is generated by the forest, losing the goods and services that come from the forest, losing the carbon storage that comes from the forest. All these have got costs to society and costs to the economy."[398]

The value of species and ecosystems that are destroyed and the resulting economic damage are several times greater over a longer period, according to Sukhdev's calculations, than the profit that can be made in the short term. With a staff of economists, biologists and social scientists, he has calculated the costs of the destruction of the forest, especially for the national economies of poorer countries. He places these costs at between 2 and 5 trillion dollars a year, when ecosystem services like the

394. Pavan Sukhdev, *Put a value on nature*, TED talk, posted December 2011, http://www.ted.com/talks/pavan_sukhdev_what_s_the_price_of_nature.html.

395. Tony Juniper, *What has Nature ever done for us*, (Santa Fe, NM: Synergetic Press, 2014).

396. Christian Gollier et al., "Declining discount rates: economic justifications and implications for long-run policy", *Working Papers*, University of Toulouse, France. http://idei.fr/doc/wp/2008/declining_discount.pdf.

397. Christian Gollier and Martin Weitzman, "How Should the Distant Future be Discounted When Discount Rates are Uncertain?", Cambridge, MA: Harvard University, November 7, 2009, informal discussion paper, http://idei.fr/doc/by/gollier/discounting_long_term.pdf.

398. Pavan Sukhdev, *Corporation 2020*, Washington, DC: Island Press, 2012.

clean drinking water that comes from forests or protections from erosion disappear.[399] These sums are real figures except for being *missing* from company or state balance sheets. A study for the British government came to the conclusion that annual additional costs of a trillion dollars may be incurred by further unchecked deforestation and the resulting increase in global warming.

As soon as the artificial walls between economy and ecology are torn down, all corporate balance sheets look very different. Fossil fuel companies that looked profitable are suddenly seen to be deeply in the red. If every hectare of rainforest yields thousands of euros per year in services for the climate, water, fruit and firewood, then business segments like the deforestation of primary tropical rainforest for palm oil plantations no longer make sense.[400] If food security and subsidies are factored in, then excessive global fishing is one long loss-making operation.[401, 402] According to an analysis by the World Bank, it causes yearly damages of 50 billion dollars.[403]

Conversely, a positive balance is created with the natural capital and new sources of profit wherever functioning ecosystems are conserved. In this way, tsunamis and hurricanes can cause significantly less damage in places where mangrove forests mitigate the effects of ocean waves. Intact forests, like those in Singapore, maintain the water supply and therefore the economic power of large cities. In countries like Madagascar, a moderate prosperity has been created in places where the forest is sustainable in the long term.[404] Environmental economists have determined that the

399. Pavan Sukhdev et al., *TEEB—The Economics of Ecosystems and Biodiversity for National and International Policy Makers—Summary: Responding to the Value of Nature*, Bonn: TEEB, 2009.

400. TEEB, *The Economics of Ecosystems and Biodiversity: Mainstreaming the Economics of Nature, a synthesis of the approach, conclusions and recommendations of TEEB*, Bonn: TEEB, 2010.

401. Boris Worm, "Impacts of Biodiversity Loss on Ocean Ecosystem Services", *Science*, Vol. 314, no. 5800, (November 3, 2006): 787–790.

402. Craig Welch, *Sea change: food for millions at risk*, The Seattle Times, December 21, 2013, http://apps.seattletimes.com/reports/sea-change/2013/sep/11/pacific-ocean-perilous-turn-overview/.

403. FAO and World Bank, *The Sunken Billions. The Economic Justification for Fisheries Reform*, Rome, Washington, DC, 2008.

404. See e.g. Thomas Elmqvist et al., "Patterns of Loss and Regeneration of Tropical Dry Forest in Madagascar: The Social Institutional Context", *PLoS One*, vol. 2, no. 5, (May 2, 2007).

careful management of forests, seen over a longer period of time, renders thirty-five times as much profit as overexploitation.[405]

While Wall Street accounts are done with daredevil virtual constructions assigning billions of "values," functioning ecosystems demonstrate a high real value, which has appeared as zero up to now on balance sheets. The extant economic system ignores "ecological profits" that benefit large numbers of people over long periods of time, while ascribing exaggeratedly high value to customary overexploitation that assists a very small, but influential group of people. What's more, governments continue to support the wrong strategies with gigantic subsidies. Sukhdev has calculated that for this reason, a trillion dollars are spent worldwide per year on environmentally harmful subsidies, such as the use of fossil fuels, overfishing or over-intensive agricultural production. On the opposite side of the sheet, significantly less funding is available for a green economy.

"Systems can be resilient up to a point, and then start a rapid decline." These words might have been used by any economist in the world to describe the world during the era of the Lehmann collapse. This sentence comes from the Sukhdev report, The Economics of Ecosystems and Biodiversity (TEEB, 2010). What might the solutions look like?

First of all, I think that it is enormously important to start in the place where future economists and managers are being trained—the MBA breeding grounds. Instead of graduating people with tunnel vision who are coached to maximize short-term profits, places of learning should take Alexander von Humboldt's holistic economic training as their role model and expand the horizons of their students rather than narrowing them. An internship at a marine research institute or in a settlement on the edge of a rainforest can hone the economic-ecological expertise that will be required in tomorrow's world.

Secondly, and long overdue, is cutting off subsidies for the destruction of the environment because they are doubly wasteful: the money promotes the wrong thing and is thus not available for the right thing.

Thirdly, accounting standards and economic indicators should be changed to reflect the whole anthropogenic reality instead of just a small segment. This includes natural capital as well as traditional monetary

405. FAO and UNEP, *Vital Forest Graphics*, Rome: FAO, 2008.

values. Whether every tree and every conservation area has to be hung with a price tag in dollars, euros or renminbi is an important issue because the monetization of nature obviously carries many risks and may have many pitfalls. Can the value of a species or habitat even be quantified? Would our view of nature become narrower and poorer if we regarded it as capital? Could it result in speculation bubbles and false incentives? Would global management consultants like McKinsey & Company have a say in how the national parks of the future should look?

There are many legitimate objections to the idea of natural capital. But it is also clear that a brand of capitalism that doesn't value the functioning basis of existence—plants, animals and people—cannot endure. Perhaps we need a phase in which we assign an economic value and a price tag to nature in order to tame the worst excesses of capitalism. Perhaps at some later point, we would rightly be ashamed that it was necessary to proceed this way. Developing a working concept for natural capital is the task for a future Nobel Prize Laureate in Economic Sciences. It's obvious that environmental damages such as carbon dioxide emissions ultimately need to carry real price tags. The atmosphere can no longer be used as a free dumping ground for greenhouse gases, and companies must not be allowed to pass the environmental damages they cause on to the community. If companies are allowed to file suits against environmental constraints in the context of free trade agreements and demand compensation through international courts of arbitration, this would be precisely the wrong direction.

Fourthly, a vital question is whether digitally accumulated money should permanently define wealth for governments and the public, or whether there are more meaningful alternatives on which governmental policies can be built. The Human Development Index of the UNDP and the Living Planet Index of the WWF, in which educational opportunities, fair distribution of wealth, environmental quality and equality are included in calculations, already express much more about peoples' real-life situations than the usual economic indices do. And this includes people like the ones I met on our trip through the border region of the Central African Republic, the Congo and Cameroon. They live from the forest— and often *in* the forest—and contribute to protection of the forest area, but their lifestyles are in danger from many sides. In a country like the

Central African Republic, where poaching and violence are so widespread, to advocate protection of the forests can endanger your life.[406] As long as it is more lucrative to sell elephant ivory on the black market than it is to conserve the forest, things will develop along negative lines. The question of what value to give natural capital is therefore particularly relevant for people in the Congo Basin. For citizens from prosperous regions, it might sound like an abstract debate but for people in the Congo, it makes an enormous difference whether forests have a value for poachers, diamond dealers and Chinese raw material companies, or whether they are protected. Many of the people we met, including a lumber mill owner, found the idea of natural capital attractive.

Meanwhile, numerous projects have been set up worldwide that economically reward the protection of forests. In the Central African Republic itself, attempts over the past several years have failed due to the unstable political situation, but in neighboring countries in the Congo Basin, some preliminary initiatives have begun. The number of field tests employing natural capital amounted globally to 170 million US dollars in 2013.[407] A program by the UN called Reducing Emission from Deforestation and Forest Degradation in Developing Countries (REDD) relies on a method whereby industrial countries compensate for their carbon dioxide emissions by paying for forest conservation programs in which carbon dioxide is bound into additional biomass.[408, 409, 410]

But these experiments have also been met with much rejection. Critics regard REDD as a modern sale of indulgences in which the rich renounce

406. Laurel Neme, "Chaos and Confusion Following Elephant Poaching in a Central African World Heritage Site," *National Geographic Online*, http://newswatch.nationalgeographic.com/2013/05/13/chaos-and-confusion-following-elephant-poaching-in-a-central-african-world-heritage-site/, retrieved Dec 2013.

407. About UN-REDD, see: http://www.un-redd.org/AboutUNREDDProgramme/tabid/102613/Default.aspx.

408. Minang, P. et al, Forestry and REDD in Africa, *Joto Afrika Journal*, no. 4, September 2010, http://www.ids.ac.uk/go/idspublication/forestry-and-redd-in-africa-joto-afrika-issue-4.

409. Rosemary Lyster et al., *Law, Tropical Forests and Carbon: The Case of REDD+*, Cambridge University Press, 2013, and Johan Eliasch et al., *Climate Change: Financing Global Forests*, London, 2008, https://www.gov.uk/government/uploads/system/uploads/attachment_data/file/228833/9780108507632.pdf.

410. Lucca Taconi, Payments for Environmental Services, *Forest Conservation and Climate Change: Livelihoods in the REDD?*, Edward Elgar Publishing, 2011.

their sins without really changing their ways. They also point to the fact that money seeps away due to corruption; that agreements ignore indigenous peoples; and that financial support is used for the wrong purposes such as fast-growing palm oil plantations that have spurious claims to binding carbon dioxide. Even more fundamental are warnings against the dangers of attaching monetary values to nature as these overlook its cultural and spiritual dimensions, and assimilate it into the capitalist utilitarian mindset and speculative way of thinking.[411]

These are all important critiques that have to be considered in a future economy of nature. But it is hard to see how some of the most important ecosystems can survive the twenty-first century without being encircled by a protective economic barrier. The old Holocene economy that relied on the great outdoors is still in full swing but its shelf life is out of date. An anthropogenic economy has to be created and perhaps better than any scientific paper on the subject, would be to test its efficacy in the form of a spectacular social experiment.

Media consumers all over the world are enthusiastic followers of shows that depict contests in which ordinary people compete against each other to prove their skills. A similar enthusiasm was evoked by the dystopian science fiction film *Hunger Games,* in which young people from a poor futuristic province had to fight each other to the death in a vast landscape until only one winner remained. Hundreds of millions of people follow these kinds of entertainment. Would their interest be equally great if the United Nations or another large organization funded a new artificial biosphere in which 100 people from all over the world moved in and tried living together until the year 2100? The *Plenty Games,* as the project might be called, would have to stretch over 1.2 square kilometers (0.46 of a mile), the approximate area of land that will be available on average to every 100 people in the year 2100. The project would resemble the *Hunger Games,* but in reverse.[412] Instead of massacring each other, the participants would have to demonstrate that they could get along with one another, survive as a community and increase both their own prosperity and the wealth of

411. Randall Abate, Climate Change and Indigenous Peoples: The Search for Legal Remedies, Edward Elgar Publishing, 2011, see also http://no-redd.com/.

412. Suzanne Colins, *Hunger Games*, New York: Scholastic Press, 2010.

their artificial ecosystems. With worldwide casting, the continents of the world could select their own representatives who would then move into the arena for the ultimate cooperation test. In the year 2100, the composition of the world participants, based on forecasts by the UN, would be broken down into the following: [413]

Asia: 43 participants
Africa: 38 participants
Europe: 6 participants
Latin America and the Caribbean: 7 participants
North America: 5 participants
Oceania: 1 participant

The *Plenty Games* could be a search for a common future, a laboratory to see how people from the most diverse cultures react to difficult situations—how they prevent them or adjust to them—and how they use and maintain their natural capital.

Africans coming from poor communities, who have endured adverse circumstances such as those in Central Africa, would live together with Americans who have left behind their SUVs and shopping malls. Europeans, with an average age of forty, would live together with young Asians and Latin Americans who would see themselves on the ascending branch of history as the natural leaders of this experiment.

It would be a tough challenge for the participants if they had to begin at the point where humankind currently stands, with the inequalities and false economic priorities of today's world. If that were to happen, the Europeans and North Americans would use a large proportion of the available resources in the first few months, much more than they actually needed and much more than the system could sustain. Based on present-day requirements, the Americans would start by introducing a turnover of 30 kilos of electronic scrap per year and per capita, 16 tons of carbon dioxide emissions, and 120 kilos of meat consumption; whereas the world average

413. United Nations, *Department of Economic and Social Affairs, Population Division, World Population Prospects: The 2012 Revision, Key Findings and Advance Tables*, Working Paper No. ESA/P/WP.227, 2013.

stands at 7 kilos, 5.2 tons and 42 kilos. They would insist on turning over 27 tons of material a year whereas an East Indian would have to make do with 4 tons.[414]

Such starting requirements would certainly not be conducive to a good atmosphere during the *Plenty Games*, especially if the Asians and Latin Americans tried to copy the Americans and the Africans continued to be excluded.

Would the groups agree right from the beginning to make a fresh start? How much of the rations and the precious first harvest would each participant receive in the first months? Would the Americans receive higher rations because of their greater body weight and hunger, and the fact that they would bring more scientists and engineers into the team? Or would the Africans receive more because they had survived on two meals a day until then? Would the Asians take on the management roles because they come from the world region with the greatest economic dynamism—or the Europeans, because they come from the home of democracy and have learned their lessons from two world wars?

What would happen if the Americans and the Europeans demanded to slaughter the chickens during the first two months to provide meat, forcing others to go without eggs? What if the Chinese secretly went fishing in the miniature sea at night? Or if a small group chopped down trees on their own initiative to make a comfortable place to sleep? In other words, what would happen if some participants acted in the same way as many people do every day in the real world: by buying themselves a personal short-term advantage that causes long-term damage to the general public and by doing so ignores natural capital?

It would be extremely interesting to observe which rules the group gives itself: for example, if they made their decisions as a community or delegated them to a select few, whether they tailored their decisions to the needs of people or preserved the well-being of the plants and animals too. It would be telling how the group reacted if natural capital dwindled; if the layer of humus soil reduced, for example, or if there was a water shortage, or the biodiversity of pollinating insects shrank.

414. Thomas Wiedmann et al., "The material footprint of nations", *PNAS online,* (September 3, 2013).

Tackled in the right way, this show could become an absolute hit across social media platforms, as audiences would experience tough conflicts and dramas as much as reconciliations and celebrations. They could watch as selfishness and community spirit collide, mentalities living in a closed system change, and what happens to people from their own cultural background when they have to live together with people from other world regions, enclosed in a small world.

Would rival groups emerge as they did during the second phase of Biosphere 2? Or would the community remain intact? Would scientific creativity thrive or would brutal survival come to the fore? Would everyone learn new skills or would some try to stash away their knowledge as a power base? Would violence break out during a famine or would the last fruit be shared? *Plenty Games* could offer some important lessons for a new anthropogenic economy because the audience would see a demonstration of how absurd it is not to integrate what we actually live off, dismissing it from our awareness as an "externality." This kind of suppression would not be possible during the *Plenty Games,* or only at the price of a rapid failure.

Provided that the participants were successful in their communal life and in the treatment of their natural capital, they should be generously rewarded. If television show participants receive up to a million dollars for knowing the name of the inventor of the pressure cooker, then role models for life in the Anthropocene should earn several times more. It is quite possible, however, that these people would forgo money after their experience on *Plenty Games* in exchange for being allowed to broadcast some of their wishes to the audience: to reduce carbon dioxide emissions, give effective aid to the poorest nations and start global ecological regeneration.

Where could such an experiment be hosted? Almost anywhere on earth where reasonably stable political conditions prevail, which unfortunately rules out Central Africa. That country is just one gigantic warning of what happens when poverty, exploitation of resources, lust for power, and indifference on the part of the world community, come together.

I have to admit that I was filled with a mixture of sadness, desperation, and relief as I climbed aboard the airplane after our exploration of Bangui and saw the darkened continent recede beneath me. Viewed from the International Space Station, much of the world appears brightly lit at night with electricity, whereas the enormous African continent remains invisible except for some dotted islands of light.

Central Africa is the counter image to *Plenty Games*: each year, it sinks further into poverty and chaos. In March 2013, yet another coup took place and intense battles between insurgents and government troops destroyed the battered capital even more. French military intervention followed.

In May, poachers invaded the saline where we had watched the forest elephants taking their mud bath. They killed twenty adult elephants and six young animals, cutting off their tusks to sell on the ivory black market.[415] For as long as ivory counts among the few treasures that the world market recognizes in a living forest, its depletion will not only continue, but will get worse.

415. "Elephant Massacre at World Heritage Site - Time to End Wildlife Crime," May 10, 2013, www.wwf.org.uk.

ELEVEN Action Potentials

Together with thousands of others in the Bella conference center on the outskirts of Copenhagen, Denmark, I had sacrificed sleep, fresh air and decent food for a week to experience this moment: to watch the most powerful politicians in the world gather on the podium with the blue carpet, laugh into the cameras, and give the thumbs-up sign. The photo, which should have been beamed across the globe to the websites of thousands of news agencies and splashed across the front cover of countless newspapers, might have borne the headline: "Copenhagen conference settles on global climate agreement and drastic CO_2 reductions."

As representatives from 200 countries, including people like John Holdren, President Obama's advisor on science and technology issues—who spelled out the catastrophic consequences of unchecked climate change—gave impassioned speeches in the conference hall. I often detoured past the podium to the corridors and back rooms in which actual negotiations took place. On no account did I want to miss that historic moment. The photo with the blue carpeted podium would have been a kind of family portrait of humanity, the evidence that *Homo sapiens* could stick together and act jointly in the face of great danger.

I had been an environmental reporter at the first UN climate conference in Berlin in 1995; was present at the gigantic global chess game in Kyoto, Japan, in 1997 when the industrial countries pledged drastic reductions in carbon dioxide; and at the UN climate conference in Bali in 2007 during which the United Nations' executive secretary on climate change, Yvo De Boer, broke down and wept. And now, in December 2008, the negotiations were supposed to be completed with a crowning success: an agreement between rich and poor nations to reduce their greenhouse gas emissions, finance the development of clean energy supplies in poor countries and decide on joint preparations for already inevitable climate change.

Hopes were high. Tens of thousands of environmental activists from all over the world had gathered in Copenhagen for the event. At his daily press briefings, Yvo de Boer summed up the opportunity to forge an agreement between the USA and China, the two greatest emitters of carbon dioxide, in the following way: "As the USA wants China to act and as China wants the USA to act, it should be possible for both China and the USA to act."

But as is well known, the Copenhagen conference faltered. The podium, prepared for a happy ending, remained empty and unused, and the final hours of the conference were worthy of Shakespearean drama.

Former allies stabbed each other in the back and fear drove parties into making irrational decisions. The superpowers accused each other of not being sufficiently prepared to act. Earlier optimism, as expressed in studies showing that reducing greenhouse gases can actually spur the economy, made way for fears of losing out on wealth.[416] The heads of government left in a hurry as it began to emerge that a substantial agreement on climate protection would fail to materialize. Even finding the smallest common denominator was stymied by a group of countries in a wild, all-night meeting at which some, like the island state of Tuvalu, tried to stave off spurious solutions, and others, like Nicaragua, used it as a way of upstaging the United States. The US promised to make as much financial support available for global climate protection as it spends in just sixteen hours on the military. Other than that, the Copenhagen conference proved to be an all-round fiasco—and one from which UN climate negotiations have not recovered, to this day.[417, 418] Steadily growing concentrations of carbon

416. Nicholas Stern, *A Blueprint for a Safer Planet: How We Can Save the World and Create Prosperity*, London: Vintage, 2010 and Eric Beinhocker et al., *The carbon productivity challenge: Curbing climate change and sustaining economic growth*, McKinsey Global Institute, 2008 and CDP, *How climate change action is giving us wealthier, healthier cities*, London, 2013, https://www.cdp.net/CDPResults/CDP-Cities-2013-Global-Report.pdf.

417. US military spending in 2010 amounted to 663.8 billion dollars; the figure agreed on in Copenhagen ran to 3.6 billion dollars for three years.

418. Nun richten sich die Hoffnungen auf den Weltklimagipfel in Paris 2015: http://www

dioxide in the atmosphere are the direct consequence. Many people now wonder whether these mega conferences are even appropriate settings in the first place to solve the problem.[419]

Climate conferences such as the one in Copenhagen in 2008 and in Paris in 2015 highlight the political and social sides of the Anthropocene. For centuries, humanity changed earth without being conscious of the global and long-term scale of its actions; now, climate conferences denote the awareness with which humanity actively and consciously influences earth. Had the conference in Copenhagen been a success, it would have vindicated the objectives of Vladimir Vernadsky, the Russian pioneer of the Anthropocene idea, who expressed his goal as uniting the ideals of democracy with the growing geological influence of humanity.[420] Vernadsky coined the term "noosphere" to describe this idea, which has been aptly translated by Andrew Revkin and Larry Kilham as "knowosphere."[421] It was not knowledge and farsightedness that were at work in Copenhagen, but rather egotism and mistrust. This illustrates the scale at which human psychology determines the physical realities of tomorrow in the Anthropocene. The psychological and social characteristics of politicians transmute into atmospheric chemistry, landscape and climate change. Negativity in the brains of politicians can result in an adverse climate for centuries to come.

Climate change is just one of many challenges of the nascent Anthropocene but perhaps it shows best the enormous political and social clashes of interest that exist between today's people: namely those who are kept happy with air conditioning and big cars, and those people of the future who might be plagued by a multitude of weather extremes; between rich nations that have built their wealth with the help of fossil energy and the

.diplomatie.gouv.fr/en/french-foreign-policy-1/sustainable-development-1097/21st-conference-of-the-parties-on/.

419. Christian Schwägerl, "UN Climate Summit Needs an Overhaul," *Spiegel Online*, December 10, 2010, http://www.spiegel.de/international/world/inside-the-chain-link-fence-un-climate-summit-needs-an-overhaul-a-734008.html.

420. Vladimir Vernadsky, *Geochemistry and the Biosphere*, Santa Fe, NM: Synergetic Press, 2007.

421. Andrew C. Revkin, "Building a 'Knowosphere,' One Cable and Campus at a Time," *New York Times, Dot Earth Blog*, January 4, 2012, http://dotearth.blogs.nytimes.com/2012/01/04/welcome-to-the-knowosphere/.

poorest nations who will suffer most from the consequences of climate change; between the old superpowers like the United States who regard their lifestyle as a God-given right and new superpowers like China who see it as their due to catch up on economic development.

Since the collapse of the Copenhagen negotiations, a question has hung in the air: *what* has to happen for something to happen? Do politicians need more terrible catastrophes like the superstorms "Sandy" and "Haiyan" before they stir themselves to action? Or is an uprising in the sinking metropolises of the South, Jakarta, Calcutta and Port-au-Prince required? Is it pointless to hope? Is the only way to stop dangerous climate change to implement a form of global eco-communism, an autocratic rule of academics whereby scientists on the IPCC decree just how much crude oil, meat and metal each person should be rationed? Or do we need active behavioral manipulation from above as was described in B.F. Skinner's *Walden Two*?[422]

These kinds of reflections are eerie, and run counter to the notion of freedom. Yet the probability of draconian measures, tough eco-taxes and duties to protect common property from present-day greed, grows with every day that too little happens. The call for authoritarian systems to relieve people of having to make the right decisions is getting louder in an unsettling way.[423] Will saving the earth from our greed require such a severe form of global crowd control? In my view, eco-totalitarian strategies would be a nightmare.

The failure of world leaders in Copenhagen shows that people should not wait for solutions from "above"—in the form of sweeping decrees by presidents and government leaders—but try to create solutions from "below," in the form of actions by individuals, families, counties, cities, companies, scientists, organizations and social movements.

The Anthropocene also poses a huge question about where power resides. A mere handful of mining companies, like BHP Billiton, Rio Tinto, Glencore, HPPL, Xstrata and Freeport McMoRan, have a disproportionate influence on the geology of humankind. They shift millions of

---

422. B.F. Skinner, *Walden Two*, Indianapolis, IN: Hackett Publishing Company, 1948.

423. Jorgen Randers, *Systematic Short-termism: Capitalism and Democracy*, http://www.2052.info/systematic-short-termism-capitalism-and-democracy/, retrieved December 2013.

tons of material and exert a covert, disproportionate, influence on political decisions such as the prevention of recycling. Exactly the same is true of fossil fuels, two-thirds of which were excavated between 1854 and 2010 by just ninety companies. The political influence of such companies, whether in the United States, Canada, Poland, Nigeria, Indonesia, China or Russia, is disconcerting.[424] Companies are not solely to blame: the responsibility is shared by their clients who demand and consume raw materials and fuels. Nonetheless, the managers and owners of these companies exert a disproportional influence over the future of earth.

It is obvious that a global climate agreement would make it easier for everyone to enforce long-term solutions as opposed to short-term economic interests. If the exorbitant subsidies for fossil fuels ended, renewable energy would immediately become relatively cheap and available to billions of people, just as ending the lavish subsidies, tax concessions and license gifts to the mining industry (4.5 billion Australian dollars annually in Australia alone) would make the reprocessing of raw materials more profitable. But due to the enormous clashes of interest between power blocs, and due to the relations of power in the global economy, where multinational companies simply buy favors from politicians, this hope might be in vain for some time to come.

Pressure on governments and companies is urgently needed; but for the time it takes for this to be successful, it might be a better strategy to act as if governments didn't exist, or as if the citizenry of the world were solely responsible for earth.

What would it mean to go about making changes as an individual?

Three pillars of strength are particularly important: the strength to achieve happiness in life from a source other than consumerism, the strength to shape the earth of tomorrow in a positive way through dynamic human communities, and the strength to gain and use new scientific and technical knowledge.

Inspiration for the mindset that helps achieve these goals comes from an unlikely source: a German aristocrat who lived in the eighteenth cen-

---

424. Suzanne Goldenberg, Just 90 companies caused two-thirds of man-made global warming emissions, November 20, 2013, *The Guardian* and Richard Heede, "Tracing anthropogenic carbon dioxide and methane emissions to fossil fuel and cement producers, 1854–2010", *Climatic Change*, vol. 122, issue 1-2, (January 2014): 229–241.

tury, known as Baron d'Holbach. Living in pre-revolutionary Paris, he belonged to a group of "encyclopedists" who tried to summarize global knowledge. In 1770, d'Holbach published a paper entitled *The System of Nature*, in which he put forward comprehensive ethical principles that could be globally valid because they were not founded on any religious basis.[425] D'Holbach expounded an early version of the theory of evolution in his paper, and he warned people about acting as if they ruled nature. He also exercised prescient environmental ethics and criticism of consumerism:

> ". . . before man covets wealth it is proper he should know how to employ it; money is only a token, a representative of happiness; to enjoy it is so to use it as to make others happy: this is the great secret…. Give the most ample treasures to the enlightened man, he will not be overwhelmed with them; if he has a capacious mind, if he has a noble soul, he will only extend more widely his benevolence; he will deserve the affection of a greater number of his fellow men; he will attract the love, he will secure the homage, of all those who surround him; he will restrain himself in his pleasures, in order that he may be enabled truly to enjoy them; he will know that money cannot re-establish a soul worn out with enjoyment; cannot give fresh elasticity to organs enfeebled by excess; cannot give fresh tension to nerves grown flaccid by abuse; cannot invigorate a body enervated by debauchery; cannot corroborate a machine, from thenceforth become incapable of sustaining him, except by the necessity of privations; he will know that the licentiousness of the voluptuary stifles pleasure in its source; that all the treasure in the world cannot renew his senses."

You can't eat money, and material wealth is not akin to happiness—that may sound like a hippie bumper sticker, quoting a Native American chief. But this thought originated in the European Enlightenment, right at the

425. Baron d'Holbach, *The System of Nature; or, The Laws of the moral and physical world*. Translated from the original French of M. de Mirabaud, London: Thomas Davison, 1820, available online http://www.philosophy-index.com/d-holbach/system-nature/i-vi.php.

time when people started to believe in progress. More recently, psychological studies have confirmed that once you have reached a rather modest level of income, personal happiness and satisfaction don't grow along with your material wealth.[426] In fact, the opposite can be the case.

But despite this, consumerism is hugely attractive for millions of people around the world. Malls are mushrooming in emerging economies, and shopping is actually a pastime for many.

Every day, seven billion people make decisions as to which of the "pleasures" of those mentioned by d'Holbach they will partake. With each purchase, they shape the world, resulting in either increases or decreases in carbon dioxide emissions, shrinkage or growth of forests and surges or falls in the numbers of factory animals. Consumption behavior already functions something like a continuous Anthropocene democracy. Every day, in millions of shops around the world, a real-time referendum takes place on the earth of the future.

The Anthropocene is the sum of the collective actions of seven, eight, nine or ten billion individuals. But individualism, at least in the Western societies that founded it, often serves as an excuse for wrong decisions. When people keep on buying bigger cars, allow themselves to use energy extravagantly or buy products from companies that treat people or the environment ruthlessly, they often have a standard excuse that runs counter to d'Holbach's imperative. They do not wonder if what they do makes other people happy, but excuse their actions by saying that what they do only has a tiny impact, if any. Many people trivialize their actions by dividing them by seven billion people. What can be problematic about something that is just 0.000 000 00014th of humanity? Surely it makes no difference if one person wastes energy or not. One in seven billion, so what? This is a standard excuse. That's how people buy stuff that is cheap now but has to be discarded after a very short time; stuff that distracts from quiet moments; stuff that optimizes short-term rewards and instant gratification. There are deep-seated processes in the body that are receptive to such rewards because they are rooted in an ancient world in which

426. Christopher Boyce et al., "Money and happiness, rank of income, not income, affects life satisfaction", *Psychological Science*, vol. 21, no. 4, (April 2010): 471–475 and Cameron Anderson, "The Local-Ladder Effect Social Status and Subjective Well-Being", *Psychological Science/* vol. 23 no. 7, (July 2012): 764–771.

rewards were few and far between. The world back then was not yet saturated with cheap sugar, cheap energy and cheap information. But today, every few hours, or even minutes, millions of people can hop from one reward to another. This system, however, is reaching its limits because it constantly has to go faster in order to keep running.

Acceleration, overdrive and stress morph into mass-scale aggression, depression, "comfort eating" and "retail therapy". Negative conditioning apparently goes so far as to make people impatient and irrational as soon as they see the McDonald's logo.[427]

In the Anthropocene, a different form of individualism is at issue in which pleasure and responsibility must remain linked together: what happens when I multiply my lifestyle by seven billion? What would happen if everyone consumed as much as I do?

Millions of people worldwide are already trying to break out of the vicious circles of consumerism and "shopaholism."[428, 429] They try to stay clear of constant instant gratification and find that there are deeper and more meaningful rewards beyond consumption. The movement of diehard car addicts, frequent fliers, meat-eaters and other "problem providers" may still have more members, but there is a counter movement growing of millions of people who are trying to learn how to live in a way that not only makes them happy, but is also good for their fellow human beings or the plants and animals in the areas where the stuff they consume actually comes from.

The best news for a good Anthropocene is that the human brain is a master at learning and relearning and that "rewiring" takes place all the time. Until old age, learning processes can cancel out old habits. Addictive and behavioral pitfalls can be replaced with new ways of living.

For decades, neuroscience told us that the brain was hard-wired and not really able to evolve after childhood. This view was based on a dogma

427. Chen-Bo-Zhong and Sanford DeVoe, "You are how you eat—fast food and impatience," Psychological Science, prepublication online, (March 19, 2010), http://pss.sagepub.com/content/21/5/619.

428. David Linden, *The Compass of Pleasures*, London: Penguin, 2012.

429. Even President George W. Bush described America's dependency on crude oil as an "addiction": see http://www.nytimes.com/2006/02/01/world/americas/01iht-state.html?pagewanted=all.

held by the founder of modern neuroscience, Santiago Ramón y Cajal, who stated in the early twentieth century that the brain matures at an early age and remains inflexible, limited and unchanged for the rest of our entire lives. But this dogma has proved to be completely untrue.[430] With more precise analytical methods, researchers are now able to see that in fact, the human brain constantly forms new synapses and cells, even into old age, while continually changing its structure under the influence of new stimuli and experiences.[431] Brain researchers' new favorite term is therefore "plasticity," that is changeability and dynamism.[432] When people change their lives, their brains also change.[433] A person who habitually hops into his car for any length of trip, but is so impressed by an article on climate change that he starts riding a bicycle and using public transport, is unintentionally changing the structure of his brain. The changes are small biochemical alterations but they can have far-reaching, tangible effects. Only a few weeks later, the rewards from cycling—better fitness, fresh air, more direct contact with other people—have become the new normal.

The plasticity of the brain puts each person in a very responsible position. Through your own actions, you can either strengthen old networks or you can stimulate the development of new nerve cells and connections. In this way, we can reinvent ourselves through our behavior and cultivate our needs and attitudes. We can either make the old paths deeper in the same way that a river carves out its bed; or by making an effort, create new paths that guide the water elsewhere. This ability to relearn, regenerate and to trust in the positive is now society's most important resource—pervading all generations, from the elderly who grew up in the economic

430. Ramón y Cajal, *Degeneration and Regeneration of the Nervous System*, Oxford University Press, 1928.

431. Steven Goldman und Fernando Nottebohm, "Neuronal production, migration, and differentiation in a vocal control nucleus of the adult female canary brain," *Proceedings of the National Academy of Sciences*, vol. 80, no. 8, (April 1, 1983): 2390–2394 and Fred Gage et al., "Neurogenesis in the adult human hippocampus," *Nature Medicine*, vol. 11, 1998, p. 1313–1317 as well as Gerd Kempermann and Dan Ehninger, "Neurogenesis in the adult hippocampus," *Cell and Tissue Research*, vol. 331, no. 1, (2008): 242–250.

432. See Mark Hübener et al., "Experience leaves a lasting structural trace in cortical circuits," *Nature*, vol. 457, (January 15, 2009): 313–317 and Ulman Lindenberger, "Comparing memory skill maintenance across the life span: preservation in adults, increase in children," *Psychology and Aging*, vol. 23, no. 2, (2008): 227–238.

433. Paul Baltes et al., *Lifespan development and the brain: The perspective of biocultural co-constructivism*, Cambridge University Press, 2006.

miracle of the 1950s and 60s, to the youngest, who are already learning about environmental problems in elementary school.

What it means to live in the Anthropocene is different for everyone depending on each person's circumstances, priorities, environment and aims. But it can be said with a fair amount of certainty that our every-day lives will have to look different if the Anthropocene is to take a more positive course. Part of this is to see through the reward mechanisms that make consumerism addictive; to be prepared to pay a suitable price for products made with care; and to become involved as a citizen in local, regional or global issues.

What would happen if 7 billion people did what I do?

My lifestyle multiplied by 7 billion =?

This is the fundamental issue of life on a planet crowded with people. In the Anthropocene, individualism means to take responsibility for the whole, in the sense that d'Holbach described. There are many things in life that remain unaffected by this multiplication and many things to which it is not usefully applied. The idea is not, of course, that all people should live in a similar or even the same way. But this calculation can protect us from many wrong actions and instruct us to be more cautious. To take our own behavior as the norm gives us the right feeling for the scale of what is happening.

It can help us develop a new sensorium for what our own lives set in motion in the world. We can try to assess how much energy and mate-rial we consume—and redesign our behaviors. We can develop an aver-sion to products that are hard to repair or difficult to recycle. We can set upper limits to our consumption of fossil fuels. We can consider the global consequences of buying industrial food and stop throwing food away.[434,][435] We can eat meat once or twice a week instead of every day.[436] We can become involved in the multitude of initiatives that nudge small, positive changes, like creating a local system of bicycle highways, looking after a

---

434. Dorothy Blair und Jeffory Sobal, "Luxus consumption—wasting food resources through overeating," *Agriculture and Human Values*, vol. 23, no. 1, (2006): 63–74.

435. Jenny Gustavsson et al., *Global losses and food waste*, (Rome: FAO, 2011) http://www.fao.org/docrep/014/mb060e/mb060e00.pdf.

436. McMichael, "Food, livestock production, energy, climate change, and health", The Lancet, vol. 370, (October 6, 2007): 1253–1263.

nature reserve or sponsoring the education of children from poorer families. And we can be open to spending money on the living world instead of dead matter: the cell phone industry grew from nothing to an industry worth billions of dollars within twenty years. So why should it be unusual to pay for conserving living communication networks in forests, marshlands and savannahs, on whose functions nature's services and our survival depend? The market for bottled water—a resource that is available in industrial countries from the faucet for a fraction of the price—adds up to 60 billion dollars. So why shouldn't we pay for carbon dioxide mitigation so that glaciers remain in place as fresh water reservoirs? It has become the norm to pay three or four dollars for a coffee—so why shouldn't it be normal to cover the costs of preserving the rainforest around coffee plantations, which, after all, increase their fertility?

The remarkable thing about consuming less and investing in good causes is that it often *increases* satisfaction. The idea behind gross domestic product (GDP) doctrine, that constant growth is necessary, is being questioned by many. Of course, our economies need to stay dynamic, innovative and competitive. A planned economy dictated by ecology in which everyone is allocated his or her portion of gasoline or bread, and which pays homage to Mother Earth in the fashion of North Korea, would be the least capable of guiding us positively through the Anthropocene. Rather, it is more about choosing the right way of life with our free will. From the individual to the state, from NGOs to multinationals, the Anthropocene is a collective learning exercise, an ongoing intelligence test for the human race.

There are already many encouraging projects, forward-thinking company managers, courageous activists, and other people who change their lifestyle in a consequential way. But for something truly new to be created, the forces of science and technology must be unleashed too. New technologies will not pave an easy path for us in the Anthropocene. We can't count on technological silver bullets that make it unimportant whether climate conferences fail and which lifestyles we choose. The hope that there might soon be a magical cooling gas, which could help us stop global warming is just as unfounded as betting on progress in chemistry as a way of solving global malnutrition.[437] Nevertheless, without a blossoming culture of

437. For an overview on this subject, see: Wilfred Rickels et al., *Large-scale intentional*

science that is made up of millions of well-educated, creative and inquisitive minds, decisive changes will fail to materialize. If climate meetings continue to break down, the responsibility of scientists to analyze the state of earth and develop new, environmentally friendly technologies will be all the greater. And for this, sufficient numbers of qualified people and resources are required.

This is why it is a cause for concern if increasing numbers of young people want to become lawyers and bankers instead of scientists and engineers. And it is even more worrisome that governments do not make enough funding available to finance scientific training and practical work. Despite some positive signals, the international network of agricultural research institutes is chronically underfinanced.[438, 439] Governments should be spending 40 to 90 billion dollars a year more on energy research according to the International Energy Agency, in order to solve the climate crisis and energy poverty in underprivileged parts of the world.[440] The cultivation of plants at public agricultural research institutions is being neglected.[441] The scientific battle against tuberculosis,[442] AIDS and

---

interventions into the climate system: Assessing the climate engineering debate, scoping report conducted on behalf of the German Federal Ministry of Education and Research (BMBF), Kiel Earth Institute, 2011; Philip Rasch et al., "An overview of geoengineering of climate using stratospheric sulphate aerosols," *Philosophical Transactions of the Royal Society A*, vol. 366, (November 13, 2008): 4007–4037; J.J. Blackstone/J. Long: "The politics of geoengineering," *Science*, vol. 327, no. 5965, (January 29, 2011): 527; Paul J. Crutzen, "Albedo enhancement by stratospheric sulfur injections: A contribution to resolve a policy dilemma?", *Climatic Change*, vol. 77, no. 3-4, (August 2006):211–219; Richard S. Lampitt et al., "Ocean fertilization—a potential means of geoengineerung?", *Philosophical Transactions of the Royal Society A*, vol. 366, (November 13. 2008): 3919–3945.

438. Natasha Gilbert, "Future funding for agricultural research uncertain," *Nature*, published online 31 March 2010, http://www.nature.com/news/2010/100331/full/news.2010.165.html and Bruce Campbell et al., "Legislating change," *Nature*, vol. 501, (September 26, 2013): 12–S14; Helen Thompson, "Food science deserves a place at the table," *Nature*, (July 12, 2012), http://www.nature.com/news/food-science-deserves-a-place-at-the-table-1.10963.

439. Jeffrey Mervis and David Malakoff, "U.S. Budget Deal Offers Researchers Some Sequester Relief", *Science*, vol. 342, no. 6165, (December 20, 2013): 1426–1427.

440. IEA Report for the Clean Energy Ministerial, *Global Gaps in clean energy R&D, Update and Recommendations for International Collaboration*, Paris, 2010, http://www.iea.org/publications/freepublications/publication/global_gaps.pdf.

441. Michael Morris et al., "The Global Need for Plant Breeding Capacity: What Roles for the Public and Private Sectors?", *Horticultural Science*, vol. 41, no.1, (February 2006).

442. Michael Morris et al., "The Global Need for Plant Breeding Capacity: What Roles for the Public and Private Sectors?", *Horticultural Science*, vol. 41, no.1, (February 2006).

malaria is predominantly being fought by the Bill & Melinda Gates Foundation but is not sufficiently financed by rich nations.[443] Even after many decades since the first warnings of dangerous global warming, the world lacks a comprehensive global network of meteorological stations: there are gaping holes, particularly in the Arctic.[444, 445] The fact that at the beginning of the Anthropocene, the majority of ecological research stations are still concentrated in classical wilderness areas but only a few are set up in the majority of the land areas where humans dominate, such as cities, agricultural areas and industrial zones, limits our view of earth in a dangerous way.[446]

If you sum up all the areas of science and technology development that are underfunded at the moment, you arrive at a staggering figure. This is also a part of the Western debt crisis and of life at the cost of the future: too little is invested in provision for the future and innovation, which amounts to the same as being a bad debt for the next generation.

As d'Holbach declared so long ago, one's own happiness is strongly linked to other people's happiness. This is even truer today, as the world is hyperconnected by the Internet and by countless means of transport. Extreme weather or drought in Africa can lead to increasing numbers of refugees trying to reach Europe, for example. Even if people are completely egotistical and only care about their own prosperity, it would be wise for them to consider the bigger picture. Retirement planning is not just about having enough money in the bank when you turn 65. If the world is in total turmoil by then, you will not have fun spending your money. So, support for green energy systems that replace the use of coal, medicines countering infectious diseases like malaria, and for agricultural

443. Mike Frick and Eleonora Jiménez-Levi, 2013 Report on Tuberculosis Research Funding Trends, 2005-2012, Treatment Action Group, New York, November 2013, http://www.treatmentactiongroup.org/sites/g/files/g450272/f/201310/TAG_TB_2013_8.5.pdf and Path, From Pipeline to Product: Malaria R&D funding needs into the next decade, Seattle, December 2013, http://www.malariavaccine.org/files/RD-report-December2013.pdf.

444. Euan Nisbet, "Earth monitoring: Cinderella science," Nature, vol. 450, (December 6, 2007): 789–790.

445. Steve Connor, Gaps in data on Arctic temperatures account for the 'pause' in global warming, The Independent, November 17, 2013, http://www.independent.co.uk/environment/climate-change/gaps-in-data-in-arctic-temperatures-account-for-the-pause-in-glo-bal-warming-8945597.html.

446. Emma Marris, Rambunctious Garden, op cit, see Footnote 233.

techniques that are suited to smallhold farmers in developing countries could be part of retirement plans in rich countries. The money moved by pension funds is already one of the most important financial forces on Earth. Why shouldn't pension funds start investing in the future of the planet? If you care about health and safety in old age, this should be of prime concern to you.

The Anthropocene confronts governments, managers, investors and individuals alike with the question of where priorities lie—in consuming what exists or building up something new; in optimizing selfish solutions or in a new global public spirit? The climate conference in Copenhagen was a sign of what can go wrong if everyone sticks to his or her own agenda and believes that this is a way to maximize his or her luck. Half a year after the conference, I listened to an audio recording of the secret backroom negotiations of the twenty-five most important government leaders.[447] The tape was leaked to the magazine I worked for at the time. By mistake, a politician had kept the microphone on that had been attached to him during a prior media interview, so a recording was made that should have stayed confidential. Journalistically, it was a real scoop that a colleague of mine gained access to this recording. What I found most interesting when preparing our report was that the words "trust" and "hypocrisy" cropped up over and over again.

Officially the talk was about carbon dioxide and fossil fuel resources, but it really was about a different kind of psychological, social and emotional resource: to lead with good example, to trust one another, to try to understand the other's perspective, to take care of those that are in a weaker position. Baron d'Holbach's text should have been read aloud to the delegates at the start of the conference.

As there is no guarantee that future climate summits will deliver any better results than the Copenhagen conference, it really is the action potential of individuals, groups and communities that counts. This is why I was happy to learn what happened in Copenhagen after the failed conference. Local politicians and citizens were so frustrated that their city

---

447. Tobias Rapp, Christian Schwägerl and Gerald Traufetter, "The Copenhagen Protocol: How China and India Sabotaged the UN Climate Summit," *Spiegel Online*, May 5, 2010, http://www.spiegel.de/international/world/the-copenhagen-protocol-how-china-and-india-sabotaged-the-un-climate-summit-a-692861.html.

had become a symbol for a failed environmental policy that they redoubled their efforts to turn Copenhagen into a green city. By 2014, they had expanded their superb system of bicycle highways so that cycling is now a dominant form of transport. The city has moved closer to meeting its energy needs with renewable energy. And throughout the city, urban farming was encouraged to produce food locally. In a world where 60 percent of people will soon live in cities, mayors and citizen groups might turn out to be more influential than the world leaders that failed to act during the Copenhagen conference. Good examples of how to live happily and wisely in the Anthropocene can spread like wildfire through social media. One cannot underestimate the action potential this creates.

CHAPTER TWELVE **Anthropocene Day**

CAN THERE BE A "GOOD ANTHROPOCENE?" No, thinks environment and science writer Elizabeth Kolbert. One should probably not use the words "good" and "Anthropocene" in sequence, she tweeted in 2014 after the *New York Times* Dot Earth blogger Andrew Revkin had posted a talk on "pathways into a good Anthropocene."

The environmental debate has been strongly influenced by negative forecasts and scenarios, like Rachel Carson's *Silent Spring*, assessments by the Club of Rome and the Worldwatch Institute or IPCC reports. Most of these warnings are solidly based on scientific knowledge and have an important role in our societies. Without them, there would be less environmental awareness and policies would be even worse than they are today. A lot of what's written in these reports cannot be negotiated politically—how carbon dioxide affects global temperature is a physical fact, not something which politicians can alter. But many of the negative forecasts share one problem: they mainly look at what will happen if current trends persist.

In his day, Robert Malthus certainly had good reasons to warn that earth could not feed more people. But he had no idea that fertilizers would be identified and synthesized from the air he was breathing and that biologists would find ways to alter the genomes of plants to make them more productive. Yes, it is still unclear whether or not Malthus will have been proven right at the end of this century, but perhaps the example of Malthus shows that it is good to think about alternatives and more positive scenarios. In the worst case, purely negative forecasts and scenarios can act like self-fulfilling prophecies: people expect the worst, lose hope that anything can be changed and start to behave cynically or carelessly. So, any report that says that "by the year 2050, this and that will happen," should really say "by the year 2050, this and that will happen *if current trends persist.*" Otherwise, a dangerous determinism will spread even further. If that happens, there is no chance people will be prepared to sacrifice time, money or short-term pleasures to change the course of events.

Yes, positive scenarios can come across as hopelessly naive and can fast become the stuff of ridicule (like so many fantasies about technofixes).[448] Or they can give the impression that problems really aren't as dangerous or as bad as we think and that the mere existence of a possible solution in the future absolves us from acting now.

But then, do we really want to be guided solely by our fears and do we really want to offer young people only negative outlooks? If collapse and catastrophes are unavoidable, "Why bother now," they might ask. If nature is doomed anyway, why love it? And if all we can do is avoid damage, aren't we just getting stuck in our own ways? For change to happen, people need something they can believe in and look forward to.

So, what follows here is a highly personal, highly speculative and extremely positive scenario of how the Anthropocene *could* progress. It's just one of a huge number of possible positive scenarios (and you don't even have to like it). But I do think it's important that there be various positive scenarios available to compete for the attention of the public, politicians, and business people. In so many of the articles that I have written, I have covered the negative possibilities; so what I have consciously omitted here are wars, the dark side of the human psyche, and the surprisingly negative consequences that can result from good intentions.[449] In sharing this vision, I am willing to take the risk of seeming like a hopeless (or rather, hopeful!) optimist, even if this might be a misconception.[450]

---

448. On the Ford Nukleon: http://www.damninteresting.com/the-atomic-automobile/; on settlements in outer space: http://settlement.arc.nasa.gov/; on robots as inhabitants of LA in the year 2013: http://documents.latimes.com/la-2013.

449. Relevant examples are plentiful including holism in nineteenth and early twentieth-century Germany (see Anne Harrington, *Reenchanted Science, Holism in German Culture from Wilhelm II to Hitler*, Princeton University Press, 1999, and the large-scale technological project of using biofuels made from farm crops to solve the energy problem (see Paul Crutzen et al., "N_2O release from agro-biofuel production negates global warming reduction by replacing fossil fuels," *Atmospheric Chemistry and Physics*, vol. 8, (January 29, 2008):389–395.

450. For a fair dose of pessimism, see my second book *11 drohende Kriege: Künftige Konflikte um Technologien, Rohstoffe, Territorien und Nahrung* (11 Looming Wars, Future

"The Scenario"

The pollution monitor in the US consulate in Shanghai is showing a new record this morning.[451] Once again, the reading is at 600, the color code is black and "beyond index." From a value as low as 50 and upward, vulnerable people can suffer from health problems. Readings of over 500 are not foreseen on the measurement scale. The Chinese government keeps broadcasting a value of 300 but inhabitants of the city don't need to go outside to know that the Americans are right, once again. Not even the fittest of adults can breathe outdoors without feeling pain. The air is a yellowish-brown mix of poisons; it stings the nose, throat and lungs. Kindergartens and schools have been closed, and even in their apartments, the citizens of Shanghai have to wear breathing masks.

In the past, when the *Asian Brown Cloud* lay over the city like a coffin lid, any form of public life was suffocated. But today, something is different. As if acting on an invisible cue, first dozens, hundreds, and then thousands of people gather in a flash mob at the Bund, the popular waterfront area on the Huangpu River. Normally, the police react fast enough to prevent this kind of unlawful assembly. But the law enforcement officers want to leave their offices due to the bad air quality. Within an hour, the crowd becomes too large to be easily dispersed. More people join in all the time. An activist who works with the Internet censorship authority had sent out an innocent-looking message—"Chinese Dream"—to all digital addresses in the greater Shanghai region. Whoever opened the message received instructions as to when to come to the Bund. Two hours later, 250,000 people had assembled there.

A young woman climbs an improvised stage and starts to denounce the causes of air pollution: corruption, short-termism, and the population's addiction to consumption. "For years, these diseases have

---

conflicts over technologies, resources, territories and food), written with Andreas Rinke, Munich: C. Bertelsmann, 2012.

451. For current readings, see http://www.stateair.net/web/post/1/4.html. The reading in Shenyen on the 22 December 2013 at 7 o'clock in the morning, for example, was 583, a value classified as "Beyond Index". For readings between 301 and 500 the following warnings apply: "Hazardous. Serious aggravation of heart or lung disease and premature mortality in persons with cardiopulmonary disease and the elderly; serious risk of respiratory effects in general population. Everyone should avoid all outdoor exertion."

been brewing in dense, dirty clouds over our heads that make it impossible for us and for our children to breathe," the woman shouts into the microphone, punctuated by coughing fits, "but I am longing for the day when our Chinese dreams appear in the sky as beautiful clouds that promise life, not death."

It's the day on which a new environmental movement is born. These are not peacenik types, nor spoiled Westerners who already have everything and are accusing the poor of wanting a part in the wealth stakes— no, these are decent, law-abiding Chinese citizens from all walks of life. At first, the official news pages try to conceal the protests from the general public. But in social media networks, images of a gigantic gathering of white dots quickly do the rounds. Because their faces can't be recognized behind the masks, people are less afraid of being caught or being followed by the police. With lightning speed, similar protests start forming in other cities that live under the *Asian Brown Cloud*, and soon many millions of people are out on the streets.

The state leaders in Beijing know that a critical point has been reached. For some time now, internal reformers have been demanding a new course. The country should no longer imitate the mistakes of the West; rather, China should take on the role of "green superpower" and develop a sustainable Chinese Dream.[452] In numerous internal meetings, the reformers have demanded that China switch completely to renewable energy sources, accept binding international targets for reducing greenhouse gases, greatly enlarge nature reserves and oblige its companies and citizens to recycle all waste.

Until now, the hardliners have had all the say. They have insisted that only continued economic growth would prevent the population from becoming dissatisfied and rebellious. There had been smog in London, Los Angeles and other Western cities for decades, they argued— it was an inevitable thing before enough prosperity had accumulated to make clean skies possible. The hard-liners talked of the day when the Chinese could live just like the Americans with big automobiles, big houses and big refrigerators. So far, the state leaders had always listened to them.

452. Peggy Liu, *China dream: a lifestyle movement with sustainability at its heart*, theguardian.com, June 13, 2012, http://www.theguardian.com/sustainable-business/china-dream-sustainable-living-behaviour-change.

There was only one occasion when the Chinese Premier faltered. A young, intelligent official, one of the green reformers' spokespeople, recalled the real strength of the ancient Empire: "China could feed itself and become strong because farmers tipped their excrement onto their fields, yielding valuable fertilizer," the young man said, "whereas today, every minute, tons of farmland are being lost through erosion, and our excrements have become toxic waste." After this incident, the Premier agreed to an experiment. A "green GDP" was calculated, that included environmental damages as economic losses. The results were catastrophic and the project was soon abandoned. Instead, hard-liners forced through greater Chinese investment in Africa in order to secure farmland.[453, 454]

With the protests spreading, the Chinese government is faced with a huge challenge. Pollution has become so bad that the country is in turmoil. The hardliners have been wrong. In astonishment, leaders in Beijing witness how protests expand beyond China's borders to other Asian regions and how the young woman who held the speech at the Bund is becoming an international star.

The Chinese protests also inspire people in the West. Young musicians in Philadelphia, who call themselves the Sexy Soils, have been experimenting with a new style of music they call bio-techno. After seeing the reports from China, they decide to release a song on YouTube earlier than planned, and immediately the clip goes viral.[455] It shows a family of four in the supermarket, but it's the parents that go nuts. They throw themselves onto the ground, screaming and thrashing their feet in the air until their children give in.

When they reach the checkout, the parents hand the bill to their kids. "Kill the Future" is an immediate hit, and a week later, a new group has sprung up that uses it as their anthem. The "panarchists" as they call themselves, walk through supermarkets taking goods out of shopping

453. Vic Li und Graeme Lang, "China's 'Green GDP' Experiment and the Struggle for Ecological Modernisation", *Journal of Contemporary Asia*, vol. 40, no. 1, (2010): 44–62.

454. Sun Xiaohua, "Call for return to green accounting," *China Daily*, April, 19, 2007 and Fergus O'Rorke, "China's revived Green GDP program still faces challenges," March 28, 2013, http://www.cleanbiz.asia/news/chinas-revived-green-gdp-program-still-faces-challenges.

455. Steffen Bauer, *Dirty, not sexy: soil is in need of attention*, Bonn: German Development Institute, December 3, 2012, http://www.die-gdi.de.

carts and putting them back on the shelves.[456, 457] This is how they get into dialogues with people about consumerism and its consequences. Only a few days after their first activities, panarchists are on the move in all Wal-Mart stores in the United States and Canada: security staff see no way of stopping them.

The general public in Europe starts to mobilize too. The European Union President expresses her solidarity with Chinese protesters and offers the government a Eurasian alliance for clean air, ecological modernization and carbon dioxide reductions. An even greater sensation is caused by TV images from the Vatican. There, the Pope goes before the press and shows reporters that he has exchanged his fine shoes for a pair of hiking boots, and his papal staff for a hiking stick. "I will leave the Vatican for an indeterminate period of time and walk around the globe, from town to town, village to village, industrial area to industrial area, and forest to forest," says the leader of the Catholic Church. "I will live among the poorest of the poor to be close to them, and knock at the doors of the most wealthy to test their hearts." The Pope's personnel distribute a hiking map showing the planned route; the journey will start from Rome in the heading towards Asia to show solidarity with the new Asian environmental movement, then through Alaska to South America where a ship will take the Pope to Africa and back to Rome. "Whether I will manage it in ten years, I cannot say," says the Pope, "but I would like to get the Church moving, and not just that: I extend an invitation to leaders of other religious communities to join me."

A few weeks later, the Chinese president gives a speech entitled "My Chinese Dream," in which he announces a new course: "Constantly better instead of constantly more." Soon after, many Republican politicians in Washington, DC feel as if their world has been turned upside down when they hear that their President, a fellow Republican, is also announc-

456. Panarchy has nothing to do with anarchy. Instead, it is about joint power over all the system's divisions. See Lance Gunderson, Crawford Stanley Holling, *Panarchy— Understanding Transformations in Human and Natural Systems*, New York: Island Press, 2002.

457. A start, if somewhat eager to please: AshEl Eldridge, "Plastic State of Mind," music video on Youtube.com.

ing a turnaround; the fear having been too great that Eurasia would turn against the United States and could leave it behind.[458, 459]

Six months after the beginning of the protests in Shanghai, leaders of the G-20 announce that "business as usual" is over. "Scientists have convinced us that in the Anthropocene, our new geological epoch, new priorities, new rules and new investments are needed to meet the challenges of our times," their communiqué states, in the usual vague fashion. During a ceremony that is broadcast worldwide, the G-20 abolishes subsidies for fossil fuels and redirects the money to a global fund for clean energy, health and education. They mandate that fundamental changes to the operating system of the world economy will be made. Central banks and the financial sector have to take long-term interests—from environmental protection to the funding of innovation—much more keenly into account. The value of the planet's "green infrastructure" and "green security system" has to be added to the balance sheets of banks and companies. The G-20 replaces GDP with a new index to measure wealth, the Earth Growth Index.[460] Banks, investment and pension funds have to introduce new criteria into their investments and bonus scheme decisions.[461]

Finally, the G-20 proposes an addendum to the UN Declaration of Human Rights: "Every individual has the right to be a part of a civilization that grows with the riches of nature and not at its cost. Every person has equal access to these riches and to the growing knowledge of humanity."

458. This is not entirely far-fetched, as William K. Reilley, Republican and former Environmental Protection Agency boss and President, George H.W. Bush, explained to me in an interview: Christian Schwägerl, "Germany Is on the Right Track: A Republican Environmentalist Finds Green Nirvana," *Spiegel Online*, April 21, 2011, http://www.spiegel.de/international/world/germany-is-on-the-right-track-a-http://www.spiegel.de/international/world/germany-is-on-the-right-track-a-republican-environmentalist-finds-green-nirvana-a-758580.html.

459. On the alternative by the name of "Chimerica" see Christian Schwägerl, "Chimerica against the world," *Spiegel Online*, December 17, 2009. http://www.spiegel.de/international/world/stalling-in-copenhagen-chimerica-against-the-world-a-667626.html.

460. There are already several attempts to set up new prosperity indicators as an alternative to GDP: for example, the Inclusive Wealth Index (http://www.ihdp.unu.edu/article/iwr), Bhutan's Gross National Happiness (http://www.grossnationalhappiness.com/) or the Better Life Index from the OECD (http://www.oecdbetterlifeindex.org/).

461. Mirjam Staub-Bisang, Public pension funds: sustainable investment pioneers, *The Guardian*, August 8, 2013,http://www.theguardian.com/sustainable-business/public-pension-funds-sustainable-investment-pioneers.

Among the resolutions is the introduction of a global upper limit for carbon dioxide emissions that lies at 2 tons per capita—the level that climate researchers consider accurate in order to prevent dangerous global warming.[462] For emissions above this limit, polluters have to pay. The revenue is given to poor countries so that they can electrify their rural communities with renewable energy.

Impressed by the positive changes, hundreds of millions of consumers start to alter their behavior and demands. Shops are forced to increase their service staff, because customers ask endless questions about how much energy and resources have been used for goods and whether manufacturers treat their workers fairly. Banks start to offer environmental and energy accounts through whereby people can monitor their environmental impact and direct funds to green projects. Major schemes are initiated to regenerate damaged ecosystems.

After a long, intensive scientific investigation, the International Union of Geological Societies announces that the human impact on Earth has been significant enough to declare a new geological epoch—the Anthropocene. "From now on, it depends chiefly on us as people as to how the Earth of the future will look. This is not only a completely new situation for geologists but for every individual," the statement reads.

It is the beginning of global ecological regeneration that leads to fundamental cultural changes. To waste energy, buy unnecessary goods and eat industrially produced meat is considered extremely uncool by young people. The last few dealers who sell gasoline-guzzling vehicles can no longer find buyers. The market for SUVs collapses. Teenagers crave robust, repairable and recyclable products, and producers adapt. Film stars in Hollywood, Bollywood and Nigeria's Nollywood set trends for more environmentally friendly lifestyles.

A Japanese billionaire buys the Tokyo fish market and transforms it into a temple to marine life. Muslim groups develop new forms of Ramadan fasting that extend to not driving automobiles. In China, cleaner air leads to a renaissance of bicycling, and in Africa, foundations are set up

462. Malte Meinshausen et al., "Greenhouse-gas emission targets for limiting global warming to 2°C," *Nature*, vol. 458, (April 30, 2009): 1158–1162, http://www.nature.com /nature/journal/v458/n7242/full/nature08017.html und IPCC Report 2013, and IPCC http://www.ipcc.ch/report/ar5/wg1/.

with the task of recovering the billions of dollars that were smuggled into Switzerland by dictators, in order to reinvest them in rural infrastructure.

On the evening news, not share prices but "Earth Growth" indicators roll across the bottom of the screen. Apps livestreaming humanity's current ecological impact are embedded everywhere. In shops, customers find fewer goods but they are significantly better in quality. This is the new source of wealth: companies focus significantly more time and resources on checking the origin of raw materials, paying for ecosystem services and designing products for long life cycles and complete recycling. They earn more on service and less on material turnover.

Anonymous discount stores have disappeared and have been replaced by "true count stores" that pay fair prices to farmers, fishermen and factory workers. Shareware and open source principles from the digital world are applied in the analog world. It becomes normal that not everyone owns a car, tools, or countless gadgets, but that these are shared.[463]

Encouraged by this, the G-20 states introduce a new round of reforms: during a summit in Washington, DC, they pass the Long Now Plan: they commit themselves to reducing their military spending to a tenth of its current amount and conversely, to a tenfold increase in spending on education, science, the environment and development.

European cities undergo facelifts. Millions of people switch to bikes, carpools and public transport. This allows parking garages to be transformed into hanging gardens where people can grow vegetables.[464, 465, 466]

463. Lisa Gansky, *The Mesh—why the future of business is sharing*, New York: Portfolio Hardcover, 2010 and Rachel Botsman, *What's Mine is Yours*, New York: Harper Business, 2010.

464. See: Kristofor Husted, "Seattle's First Urban Food Forest Will Be Open To Foragers," *NPR News*, March 1, 2012, http://www.npr.org/blogs/thesalt/2012/02/29/147668557/seattles-first-urban-food-forest-will-be-free-to-forage; David Tracey, *Urban Agriculture: Ideas and Designs for the New Food*.

465. Korea:http://ecomobilityfestival.org/;Bogota:http://www.ifhp.org/ifhp-blog/mobility-bogot%C3%A1; Copenhagen: http://copenhagenize.eu/; London: http://www.london.gov.uk/sites/default/files/Cycling%20Vision%20GLA%20template%20FINAL.pdf; San Francisco: http://www.sfgate.com/bayarea/article/San-Francisco-bicycle-boom-follows-bike-friendly-5060338.php#photo-5591168.

466. Some car makers already develop integrated systems where cars are produced with renewable energy, can run on renewable energy and are integral part of a network of buses, bikes and other means of transport.

Large parking lots and traffic tailbacks are deemed "totally twentieth century." Garbage is seen as "very Holocene."

Trailblazing cities around the world convert highways into greenways. There is plenty of space for bicycles, buses and car poolers, but otherwise the asphalt is removed to bring the earth to light again. Where parking lots used to be, hedges and flowerbeds, playgrounds, outdoor offices and cafés now appear.

In China the first aerial farms are built—high up on skyscrapers, aquariums are linked to greenhouses to create closed nutrient cycles. Insect breeding is integrated into these farms and augments the protein supply of the inhabitants.[467]

Out on the high seas, things are changing, too. The United Nations agree to hire thousands of police and marine biologists to protect the oceans from over-fishing and pollution.[468] Drones are equipped with gigantic nets to capture trawlers fishing illegally and haul them into the next port. Half of the world's oceans are turned into International Marine Parks, with the effect that fish stocks in the other half multiply, so coastal communities around the globe thrive again. Talking about the *merroir* of fish becomes fashionable for food lovers.[469, 470]

At around the same time, the color green takes on a new meaning in Arab democracies. Baghdad is no longer referred to as the capital of terrorism, but instead as a place that attempts to build on its past heyday, between the eighth and eleventh centuries, and re-establish itself as the global Harvard or MIT. Young people from Casablanca to Jakarta become proud of the fact that Islamic scholars were the ones who developed algorithms and gave the world numerous insights into astronomy, chemistry and medicine.[471]

467. See: Turning urban infrastructure into Art, http://www.fastcoexist.com/3017462/turning-urban-infrastructure-into-art?utm_source=facebook#2.

468. What can happen in this case is shown in this wonderful film "Play Again," http://playagainfilm.com/.

469. Taras Grescoe, *Bottomfeeder: How to Eat Ethically in a World*, New York: Bloomsbury, USA, 2008.

470. German Advisory Council on Global Change (WBGU), Governing the Marine Heritage, Berlin 2013, http://www.wbgu.de/en/flagship-reports/fr-2013-oceans/.

471. For further reading, see: Jim al-Khalili, *The House of Wisdom: How Arabic Science Saved Ancient Knowledge and Gave Us the Renaissance*, Penguin, 2012.

Inspired by the panarchists, a powerful movement called the Green Crescent sweeps the Islamic world with the aim of resurrecting the cedar forests and fertile farmland that once stretched across North Africa and the Middle East. The wounds of erosion and salination soon begin to heal.[472]

Global agriculture undergoes a series of astonishing changes. All genetic information becomes freely available in open source databases. Monocultures are superseded by fields with many different crops. Trees and hedges grow on farmland to increase productivity. Botanists and indigenous people team up and soon after, many hundreds of new types of fruit and vegetables hit the grocery shelves. Traditional knowledge is combined with modern biotechnology so that pesticides become a thing of the past. Farmers now see themselves as ecological network administrators. Besides producing food, they get paid to look after pollinating insects, soil microbes and bird diversity.[473] Pupils regularly spend several weeks at a time on special school farms where they are taught basic skills in everything from hunting to food production and biotechnical practices.

When farmers begin to grow bioadaptive materials—such as basic material for compostable automobiles—a global rural renaissance sets in. Well-educated parents and their children move back to the countryside, bringing with them 3D printers with which they are able to manufacture all the tools, basic commodities, and electronic goods they need. "Regeneration food" comes into fashion, which originates from degraded land yet contributes to its recovery. In industrial wastelands tailor-made plants and microbes extract toxic contaminants from the ground.[474]

The rise of atmospheric greenhouse concentrations first slows, then stalls. After scientists develop microorganisms and nanobiological catalyzers that extract carbon dioxide from the air and turn it into raw material for industrial use and construction, peak carbon dioxide is reached.[475]

472. Rattan Lal et al., *Recarbonisation of the Biosphere—Ecosystems and the Global Carbon Cycle*, Berlin: Springer, 2012.

473. Austria is one of the countries that already works with farmers to monitor biodiversity changes: http://www.biodiversitaetsmonitoring.at/.

474. For a research project in biomining, see http://www.helmholtz.de/en/artikel/biomining-metal-extraction-with-bacteria-1788/.

475. Peter Styring and Daan Jansen, *Carbon Capture and Utilisation in the green economy*, The Centre for Low Carbon Futures, 2011.

At this time, global regeneration starts to be reflected in language too. Words, such as "waste" and "environment" slowly disappear. They are replaced with new terms such as "flowstuff" and "invironment" Derived from ecology plus economy to denote the study and management of sustainable living, a new word—"ecolomy"—starts circulating, and finally becomes part of everyday vocabulary. Previously industrial nations change their names to "regenerative nations" and former developing countries call themselves "bioadaptive countries". Red Lists are replaced by Green lists of regenerating populations of formerly endangered animals and plants.

Due to these changes, the world in general is more relaxed. There are fewer military conflicts, less environmental stress and significantly fewer deaths from infectious diseases.[476] Research into bioadaptation makes huge strides. Scientists at the Indian Institute for Technology in Chennai construct biomimetic submarines—machines with organismic characteristics that feed on marine plastic waste.[477]

Harvard researchers develop houses that heat and cool themselves with artificial fur and capillaries in their walls. Facades change color using nanobiological sensors, quite like octopi.[478] At Berlin's Natural History Museum, researchers make progress with "aerobionicles, tiny sensors that can monitor the health and diversity of ecosystems.[479]

The planet is covered with a dense system of sensors: satellites, drones and monitoring stations form an "Internet of Life."[480] People soon say that they "live *in* Earth" instead of "*on* Earth."

476. For the link between the environmental crisis and military conflict, see among others UNEP Year Book 2010 at www.unep.org/yearbook/2010; Tom Spencer et al., *Climate Change & The Military: The State of the Debate*, prepared for the IES Military Advisory Council, December 2009, http://www.envirosecurity.org/cctm and German Advisory Council on Global Change, World in Transition: Climate Change as a Security Risk, Flagship Report 2007, Berlin, 2007, http://www.wbgu.de/fileadmin/templates/dateien/veroeffentlichungen/hauptgutachten/jg2007/wbgu_jg2007_engl.pdf.

477. Critical analysis by Manuel Maqueda ar http://kumu.cc/2013/03/27/those-crazy-plastic-cleaning-machines/ but his objections can be overcome if these machines are self-sufficient. It is better, of course, to reduce plastic pollution at source or to recycle plastic in social projects. See: http://plasticbank.org/.

478. In Hamburg a building with a façade made of algae was built in 2013: see Taz Loomans, "The World's First Algae-Powered Building Opens in Hamburg," April 14, 2013, http://inhabitat.com/the-worlds-first-algae-powered-building-opens-in-hamburg.

479. http://conservationdrones.org/.

480. For a depiction of the current worldwide network of environmental monitoring systems, see: http://thingful.net/.

Across the globe, the date when the masses came together in Shanghai is celebrated annual as "Anthropocene Day". Every ten years, a great festival takes place; on which occasion, the group of roaming religious leaders completes a tour of earth and comes to Rome for a month to rest. Every 100 years, people celebrate the Feast of the Long Now that lasts an entire month, and whose customary curtain-raiser is the classic bio-techno sound of the Sexy Soils.

EPILOGUE Deep Future

W E HAVE TRAVELED A LONG WAY from the atoms that conglomerated to form planet Earth to the origin of life; from the first organisms that shaped the course of evolution to early hominids; from the foundations of our species to the advent of agriculture, cities, and technology; from early religious thinkers to our complex pluralistic theorists of today; from a period when nature seemed boundless to a present when everything comes back and there is no outside anymore; from environmental alarm to the question of what can and should be done; from doomsday fears to an alternate possible course of events.

Welcome to the Anthropocene, the epoch in which our thoughts turn into geology and our consciousness pervades nature (and vice versa). That's what Teilhard de Chardin, the Jesuit geologist, called *complexification* and it's easy to see why he coined that term.

Complexification is all about increasing freedom. The more different states a system can reach, the greater the number of choices made along the way. The Anthropocene is not one predetermined thing. It's a whole new universe of possibilities and choices. There is no Anthropocene doctrine, no predestined outcome, no fixed goal. It's an open-source, open-end, open-everything process (apart from the physical, geological and biological laws by which we are bound).

Even Paul Crutzen, the man who (in conjunction with Eugene Stoermer) coined and popularized the Anthropocene concept, has no idée fixe as to what it all means. I picked him up from the airport once in Berlin when he came to attend a conference within "The Anthropocene Project." One of the first things he said to me was: "The Anthropocene, what is it, really? Nobody yet knows."

A while later, Paul Crutzen sat at my kitchen table, where we enjoyed a light snack. We had been talking for two hours about his scientific career and were starting to edge towards the future. This eighty-year-old man—a great intellect embodied in a small frame—was letting me take him on a journey through time. I was interested in whether he thought that humanity will use geoengineering in the future; in other words, whether technical measures will be used to fight climate change, by injecting special gases into the atmosphere. It has often been said that Crutzen is promoting geoengineering and it is not unusual for people to believe that geoengineering and the Anthropocene idea are identical.[481, 482] But Crutzen's answer was unambiguous:

> *"I would not apply geoengineering at present, but research should be carried out and I'm already surprised by how many pursue it as a research topic. I share the fear, however, that researching geoengineering will lead to an attitude that carbon dioxide reductions can be postponed because sulphur injection technology will save us from dangerous climate change. That would be totally wrong. I am doubtful that geoengineering will be used because of its cost and its side effects. We should definitely not count on it."*

There is no quick and easy way out of the looming climate crisis; no magic pill that humanity simply has to swallow to be cured. Crutzen knew all the evaluations carried out in this area because he'd either done them himself or he'd read them. This made him wary of "technofixes"—the idea that a small caste of engineers will solve the problems for the rest of us without our even realizing it. In Crutzen's opinion, there was no way around drastic reductions in carbon dioxide emissions, which concerns nearly every individual. But I wanted him to expand the horizon of time a little beyond the twenty-first and twenty-second centuries. When societies learn to

---

481. Eli Kintisch, "The Geopolitics of Geoengineering," *MIT Technology Review*, December 17, 2013, http://www.technologyreview.com/review/522676/the-geopolitics-of-geo -engineering/.

482. See also the outstanding book by Eli Kintisch, *Hack the Planet*, Hoboken, NJ: Wiley 2010; Corner and N. Pidgeon, "Geoengineering the climate: the social and ethical implications," *Environment*, vol. 52, (January-February 2010): 24–37, 2010 and the Report Governing Geoengineering Research: A Political and Technical Vulnerability Analysis of Potential Near-Term Options, RAND/TR-846-RC, 2011.

think on the geological scale, perhaps attitudes towards climate change could alter in surprising ways?

> Me: *When people start thinking on a huge geological timescale, might that lead to making today's problems look smaller?*
>
> Crutzen: *This is a very important point that needs further debate. How long are the time-scales on which we can think and act? What we have is that scientists and engineers add to earth's knowledge pool, year by year, in a catalytic fashion.*
>
> Me: *For instance one could say that climate change might be bad now but that it will stop the next Ice Age and thus be a good thing in the long term . . .*
>
> Crutzen: *Through most of human history, much wider parts of earth were covered with ice than today. But now our civilization is tuned to the post-glacial climate in which it thrived. If another Ice Age was starting, our descendants would probably do everything in their might to stop it. And that may actually be OK. But I don't think one can justify today's global warming with the argument that it will stop another Ice Age. Global warming probably will play out on such a long time scale, but we are not yet in a position to make conscious and educated decisions about questions like this.*

Once again, Crutzen was extremely cautious. His experience with chlorofluorocarbons had made him humble in the face of earth's complexity. He did not suffer from delusions of grandeur that humans can control earth like an engineer controls a machine. I wanted to know more about what he expects from the Anthropocene.

> Me: *If a fictitious geologist examined the legacy we leave behind from our epoch, one million years from now . . . ?*
>
> Crutzen: *One million years? You assume that there is intelligent life around to collect and interpret the signals. That is a very optimistic assumption, looking at the situation right now.*

In the blink of an eye, we had traveled a million years into the future in our conversation—a very Anthropocene thing to do. I thought he would tell me something about fossil traces from our epoch, about petrified automobiles, plastic leftovers and carbon dioxide bubbles in the ice. But Paul Crutzen didn't share my assumptions. He was not comfortable with the idea of a fictitious geologist in the distant future who walked around wearing the clothes and carrying the tools typical of today, because in a million years' time, of course, there will be no geologists as they exist today. I understand that. On the other hand, the fact that he thought it unlikely that intelligent life will even exist in the distant future surprised me. Wasn't that exaggeratedly pessimistic? If a fictitious biologist had asked himself a million years ago about conscious life, he would probably never have guessed that there would be Nobel laureates—and yet here we were. I knew that in the course of his research into chlorofluorocarbons and the topic of a nuclear winter, Paul Crutzen had intensively explored horrific scenarios, some of which included the wiping out of intelligent human life. This surely shapes a person's outlook. But speculations like mine over such long periods are completely imaginary and free, and so I thought that perhaps a fundamentally pessimistic attitude was showing through.

Me: *Have you remained an optimist?*

Crutzen:*Did I say I was an optimist?*

He looked at me challengingly. When a PhD student once asked him which research questions were currently the most pressing, he replied: "Even if I knew that, I wouldn't tell you because you have to find that out for yourself." His answer to me was similarly abrupt. He realized that I was hoping for a more positive reply and smiled. In fact I'd thought: What would young readers think when they read these lines? Would they think that the Anthropocene was just the sum of all environmental problems? That would contradict what Paul Crutzen wrote five years ago—that after industrialization and the Great Acceleration, the Anthropocene might be the era when people became "stewards of the earth."[483] I didn't want these to be the final words of our interview.

Me: *Don't you hold out any optimism for the future in the Anthropocene?*

483. Will Steffen, Paul J. Crutzen and John R. McNeill, "The Anthropocene: Are Humans Now Overwhelming the Great Forces of Nature?" op. cit., see Footnote 290.

Crutzen: *Of course, of course, there are many things that make me feel positive, most of all the creative strength that can be found in art and literature.*

So there was hope after all—in art and literature.[484] It is interesting that this man, of all people, who had often been falsely depicted as an advocate of ruthless geoengineering, did not cite scientists or technologists as his first source of hope. In Paul Crutzen's Anthropocene, creativity, empathy and diversity of perspectives are the most important features. This was no stereotypical "male white scientist" before me who wanted to condescend with his explanation of how the world works. Crutzen does not have any preconceived ideas of what the Anthropocene must or must not be. On his many research expeditions, while he collected air samples in high-tech lead canisters from burning rain forests, smog-polluted cities and polar areas, he encountered many different cultures and knew that there was more than one point of view. But can creativity really be the sole reason for hope, I wondered.

Me: *And apart from that?*

Crutzen: *What makes me feel positive is that through, and especially due to negative effects, we understand the world better. My research work on how vulnerable the atmosphere is made me very fearful. But since then I have comforted myself with the thought that environmental problems in particular have led to the efforts to understand how our environment actually works.*

Paul Crutzen saw that this answer had caught me completely by surprise and he smiled again. Negative effects, fear, environmental problems—his words sounded negative. But in truth, they pulsed with an optimism that

484. For contemporary artistic interpretations of the Anthropocene, see for example the works of Yesenia Thibault-Picazo (http://yeseniatp.com/), David Thomas Smith (http://cargocollective.com/dtsmith/ANTHROPOCENE), Edward Burtynsky (http://www.edwardburtynsky.com/), David Schnell (http://www.eigen-art.com/index.php?article_id=92&clang=1), Vincent Fournier (http://www.vincentfournier.co.uk/site/), Natalie Jeremijenko (http://www.nytimes.com/2013/06/30/magazine/the-artist-who-talks-with-the-fishes.html?_r=0 and http://www.ted.com/talks/natalie_jeremijenko_the_art_of_the_eco_mindshift.html), Jens Harder (http://www.hardercomics.de/) and J. Henry Fair (http://jhenryfair.com/aerial/).

is much greater than if he'd just said that humanity will somehow get its act together, and isn't likely to end up in the geological gutter anytime soon. The positive grows from the negative—this is what Crutzen experienced at first hand when he discovered the dangers threatening the ozone layer, and global politics reacted soon afterward with a ban on chlorofluorocarbons. But he was talking about something even deeper; he was saying that we get to know our environment better by polluting it. This is a dialectical statement that packs a punch. It allows us to hope that what we refer to nowadays as "environmental problems" will not distance and alienate us from the earth but have quite the opposite effect in that they will bring us closer to the earth and help us grow together. There is an echo of Hölderlin's maxim that where danger threatens, the power of salvation also grows.

This statement is pessimistic and optimistic at the same time, and it builds a bridge into a long future for humanity on earth.[485] It opens up a world in which the monstrous effects of all the things that Paul Crutzen listed in 2002 in his "Geology of Mankind" article, link us more deeply to the earth and sharpen our urge to understand what we are doing.

No one knows what the future of *Homo sapiens* holds. We are still a very young species that has only been around for approximately 250,000 years. Is it half-time, will we exist for another 250,000 years[486] or only another 25,000, or much longer—perhaps another 5 million years? Just consider the Coelacanth fish *Latimeria*, a so-called living fossil that has inhabited earth for 400 million years.

The entire planet is going through a kind of human bottleneck: at the very least, the changes that humans have wrought will impact all future patterns of development here. What can be predicted with equal certainty is that sometime in the future, the Anthropocene will be over and non-human processes will once more win the upper hand. Advanced calculations predict that in 4 billion years, our Milky Way will collide with

485. For the long version of the interview, see Nina Möllers and Christian Schwägerl, "Welcome to the Anthropocene—Our Responsibility for the Future of Earth", Munich: Deutsches Museum, 2014.

486. Curt Stager, *Deep Future—the next 100,000 years of Life on Earth*, New York: Thomas Dunne Books, 2011.

the Andromeda nebula, and the intensity of the sun will critically decline in a few billion years before our star expands in one last surge.[487, 488] In his book *The System of Nature*, Baron d'Holbach speculated back in 1770 that "the human species is a production peculiar to our sphere, in the position in which it is found: that when this position may happen to change, the human species will, of consequence, either be changed or will be obliged to disappear."[489] But this only means that for the next 100 million, if not billion, years ahead, earth will be able to accommodate life.

What happens between now and the distant future remains completely open. Perhaps birds will sing in the style of Johann Sebastian Bach's compositions. Mainstream science fiction has explored many possibilities, from ultra-intelligent humans to a future of degenerate apes.[490]

Perhaps one day, a new intelligent life form will care in the same tender way for the last human being as humans care nowadays for river dolphins and blue whales.

In the Anthropocene, we are entering a process that changes the conditions, bases and strengths of our further evolution and we are entering into new symbioses with plants, microbes, fungi and animals but also with members of the technosphere, such as sensors, robots and databases.[491]

In the midst of widespread destruction, pollution, violence and disorientation, the characteristics of a new civilization are becoming visible. If creative and protective activities win the upper hand over destructive processes, then the biological-technical networks of the "knowosphere" could evolve into a kind of collective consciousness of the planet, a 4D Inter-

487. NASA model of the collision with the Andromeda nebula: http://www.nasa.gov /mission_pages/hubble/science/milky-way-collide.html.

488. Jason Palmer, "Hope dims that Earth will survive Sun's death," *New Scientist*, February 22, 2008, http://www.newscientist.com/article/dn13369-hope-dims-that-earth -will-survive-suns-death.html.

489. Baron d'Holbach, *The System of Nature* or, *The Laws of the moral and physical world*, op. cit., see Footnote 423.

490. An inkling of these can be found in the futuristic videos by the artist Stefan Larsson, for example, "Circular Confabulation" (http://vimeo.com/18295577) und KIIA (http://vimeo.com/6796931).

491. Michel Houellebecq, *The Possibility of an Island*, London: Vintage International, 2007; Kurt Vonnegut, *Galápagos*, New York: The Dial Press, 1999; Steven Spielberg, *Artificial Intelligence A.I.*, Amblin Entertainment, 2001.

net not only of things, but of all organisms; an Internet of Life or a slowly emerging form of Gaia. With the help of humans, technology—which is really transformed geology—might develop into a system that connects life rather than destroys it. D'Holbach also foresaw this in 1770: "Let man elevate himself by his thoughts above the globe he inhabits, then he will look upon his own species with the same eyes he does look upon all other beings in Nature." One might argue that in order to achieve this paradigm shift, it might be better to name our geological epoch the "Biocene" to reflect the symbiotic nature of life on Earth.

For the time being, it seems more important to express in geological terms that humans have become the dominant force of change on Earth and that it is high time for us to take full responsibility for our actions. From this acknowledgment and its attendant care and respect, a symbiotic way of living can develop. To perceive oneself as acting on the geological scale might understandably be a little too daunting and weighty for some. And considering the time it takes to confront and solve problems like climate change, malaria and biodiversity loss, it can be hard to remain confident.

Then it helps to have Paul Crutzen's comforting statement: that our understanding of earth increases with the increase of environmental problems. The big question of course is which grows faster.

Perhaps it helps to reflect how long it took (and is still taking) humankind to ban slavery, to give women equal rights, to accept universal human rights, to create a system of international cooperation, to install democracy. One hundred years ago, people believed they were technologically advanced but there was no Internet, no human genome project, and no smart phones. People speculated about the future in the same way as we do now but they didn't imagine how we live today. Our lives would have been pure science fiction for them.

At my kitchen table, Paul Crutzen has a hopeful message: "We are not doomed."

Our species is full of potential. This is what forerunners of the Anthropocene expressed when they described the *noosphere*—the sphere of human intellect and knowledge—as a fusion of democracy and geology. Earth is still a wonderfully rich planet with endless possibilities. And unless Newton's laws propel another celestial body, a second Theia, into our planet, we have a long time to explore it, understand it and shape it. We are not living at the end of the world but in the middle of our planet's lifetime, and perhaps even in the middle of our own species' maturation process, an exit from puberty on the scale of a planet, as Andrew C. Revkin has put it, when "those remarkable human traits, self-awareness and empathy catch up with potency."[492]

492. Andrew C. Revkin, "Puberty on the Scale of a Planet." *New York Times*, Dot Earth, August 7, 2009.

ACKNOWLEDGMENTS

Many great friends, family members and colleagues have supported me in my life and work and my projects over the years and in becoming able to write a book about the global phenomenon called the Anthropocene. An exhaustive list would name hundreds of wonderful minds and hearts, so let me just highlight a few: My parents have offered me the chance to roam around and explore nature since my childhood, for which I am very grateful. Colin Clubbe and Deborah Seddon have helped me develop my biological skills while living in England and have been dear friends ever since. Steven Ware and Seymourina Cruse have been a huge source of inspiration on many levels. Fred and Kay Wolff have helped me to free my mind in the Californian desert and have been early supporters of exploring the Anthropocene. Frank Schirrmacher and Joachim Müller-Jung have been great supporters during my wonderful years at *Frankfurter Allgemeine Zeitung*. Without my agent Matthias Landwehr, this book would not exist. Reinhold Leinfelder and Helmuth Trischler were great company in exploring the Anthropocene after the German edition of this book came out in 2010. Achim Steiner and Paul J. Crutzen shared a lot of their insights and wisdom with me.

Special thanks are due to the German translator, Lucy Renner Jones, and the editor at Synergetic Press, Linda Sperling, for their excellent care in translating and editing this updated English edition.

Finally, a word of thanks to the birds that keep showing up in the shrub a few meters away from my desk and who allowed me to marvel at them and regain strength and joy while writing.

Christian Schwägerl, Berlin, 2014

GLOSSARY

Aerobionicles Conservation drones that have tiny sensors that monitor the health and diversity of ecosystems; the term is derived from a line of futurist Lego toys.

Anthromes Globally significant ecological patterns created by sustained interactions between humans and ecosystems, also known as anthropogenic or human biomes.

Anthropoids Ape-like creatures resembling humans.

Apocalypticism Belief that the world will end very soon due to a catastrophic global event.

Archean Geologic period before the Proterozoic eon, 2.5 billion years ago.

Asian Brown Cloud Layer of air pollution caused by the burning of fossil fuels and biomass that reoccurs from November through May, mostly over India and Pakistan.

Australopithecus Genus of hominids that evolved in East Africa 4 million years ago, becoming extinct 2 million years ago.

Aymara An indigenous people, about 2 million strong, who live in the Andes Mountains and Altiplano regions of Bolivia, Peru and Chile.

Bioadaptation As distinct from external visible physical adaptation to the biophere, bioadaptation is a neologism for organic molecular structural alteration.

Biocides Chemical agents or microorganisms that can destroy or render harmless an organism; commonly used in medicine, agriculture and industry.

Biomass Plant or plant-based material grown and collected for use as fuel.

Biome Ecosystem extending over a large area, containing distinctive communities of plants, animals, and soils.

Bionauts Russian neologism for scientists working in closed ecological systems, derived from Greek: *bíos* (life) and *naut* (sailor).

Biosphere The layer of organic life on earth, integrating all living species, which is thought to have arisen about 3.5 billion years ago.

Biosphere 2 Materially closed ecological system, situated in Oracle, near Tucson, Arizona, designed to explore interactions between humans, agriculture, technology and nature.

Cambrian Era   Geological period of the Paleozoic Era, from 540 million to 485 million years ago, succeeded by the Ordovician.

Carbon cycle   Biogeochemical cycle by which carbon atoms are recirculated throughout the biosphere, comprising a sequence of events key to making earth capable of sustaining life.

Cenozoic   Current and most recent of three Phanerozoic geological eras, covering the period from 66 million years ago to the present and popularly termed the "Age of Mammals."

Chlorofluorocarbons (CFCs)   Organic chemical compounds containing carbon, chlorine and fluorine, widely used as refrigerants, propellants and solvents. Manufacture of CFCs has been phased out under the Montreal Protocol because they cause ozone depletion in the upper atmosphere.

Cold War   Term used to describe the political and military tension between Western powers and the USSR, from 1947 to 1991. It was "cold" because no actual battles were directly fought between the two sides.

Columbian Exchange   Widespread exchange of animals, plants and diseases between America and Eurasia, following Columbus's voyage in 1492, which circulated a variety of new crops, like maize, potatoes and tomatoes.

Cyanobacteria   Bacterium that obtains its energy through photosynthesis, absorbing carbon dioxide and releasing oxygen.

Diatoms   A major group of algae; one of most common types of phytoplankton.

DNA sequencing   Process of determining the exact order of the four nucleotide bases—adenine, guanine, cytosine, and thymine—within a strand of DNA.

Earth Modeling   An applied science creating computerized representations of portions of the earth's crust, based on geophysical and geological observations made on and below its surface.

Einkorn wheat   "Single grain," referring to the wild species of wheat, *Triticum boeoticum*, or its domesticated form, *Triticum monococcum*.

El Niño   Warm ocean current of variable intensity that periodically develops off the Pacific coast of South America, causing unusual weather patterns.

Emmer wheat   Grain that was one of the first crops domesticated in the Near East's Fertile Crescent.

Eosimias   Genus of early primates, identified in 1999, meaning "dawn monkey," that lived 40 to 45 million years ago and was likely a tree dweller, reliant on a diet of insects and nectar.

Fertile Crescent   Region containing moist and fertile land in and around the Tigris and Euphrates rivers.

Food Chain   A linear sequence beginning with plant-life and ending with animal-life. A simple food chain starts with grass, which is eaten by rabbits that are, in turn, eaten by foxes.

Fossil Fuels   Combustible fuels (oil or coal) formed by natural anaerobic decomposition of dead organisms, laid down millions of years ago.

Freon®   Registered trademark held by DuPont Corporation, for a number of products that have typically been used as refrigerants and aerosol propellants, including chlorofluorocarbons (CFCs), implicated in ozone depletion.

Genetic engineering   Direct manipulation of an organism's genome; new DNA is inserted into the host genome. Genes may also be removed using a nuclease (enzyme that cleaves the chains of nucleotides in nucleic acids into smaller units).

Geochemical Cycles   Pathways that chemical elements take on the surface and crust of the earth, by means of subduction and volcanism.

Geosphere   Collective name for the lithosphere, hydrosphere, cryosphere and atmosphere.

Grafenwöhr   Town in eastern Bavaria, Germany, known for its large US Army installation and training area nearby.

Great Pacific Ocean Garbage Patch   A gyre (system of rotating ocean currents) containing man-made debris, in the Pacific Ocean, characterized by exceptionally high concentrations of plastics and chemical sludge trapped by the currents.

Green Revolution   Term first used in 1969 to describe modern agricultural developments that increased food production worldwide, including high-yield grains, irrigation, hybridized seeds, fertilizers and pesticides.

Greenhouse Gases   Gases emitted into the atmosphere that contribute to global warming, including carbon dioxide, methane and nitrous oxide.

GSSP   Global Boundary Stratotype Section and Point; an internationally agreed upon reference point on a stratigraphic section which defines the lower boundary of a stage on the geologic time scale.

Haber-Bosch process   Industrial process combining nitrogen and hydrogen to produce ammonia, used to make fertilizer.

Holocene   Geological epoch meaning "entirely recent," which began at the end of the Pleistocene and continues to the present day.

Hominid   Genera also known as great apes. Hominid species other than Homo sapiens are extinct.

Hominoid   Apes. Except for gorillas and humans, hominoids are agile climbers of trees. Most non-human hominoids are rare or endangered due to loss of habitat.

Homo sapiens   Scientific name for the present day human species.

Humanoid   Resembling a human being.

International Space Station   A habitable artificial satellite, in low Earth orbit.

Internet   Not to be confused with the World Wide Web, the Internet is a global system of interconnected computers that carries an extensive range of information resources and services.

IPCC   Intergovernmental Panel on Climate Change, a scientific body under United Nations auspices, which produces reports that support the United Nations Framework Convention on Climate Change (UNFCCC).

Iron Curtain   Term used by British Prime Minister Winston Churchill to describe the fortified border between West European and Soviet controlled East European countries after World War II.

Jurassic   Geologic period from 201 to 145 million years ago, named after the Jura Mountains in France, where limestone strata from the period were first identified.

Kamoyapithecus   African primate of the late Oligocene period, 24–27 million years ago.

Limnologist   Scientist who studies freshwater inland waters.

Magma   Mixture of molten or semi-molten rock found beneath the surface of the earth.

Maha-Kalpa   Sanskrit term for a huge period of time, thought to be about 3.42 billion years.

Megacity   Metropolitan area with a total population in excess of ten million people.

Methane   The main component of natural gas. Its relative abundance makes it an attractive fuel but atmospheric methane is a potent greenhouse gas.

Monoculture   Modern industrial agricultural practice of growing a single crop or plant over a wide area, year after year, which can lead to accelerated spread of pests and diseases.

Montreal Protocol   International treaty designed to protect the atmospheric ozone layer by phasing out damaging airborne chemicals. Since the treaty was ratified in 1989, the ozone "hole" over Antarctica is "recovering."

MRT   Magnetic resonance tomography (MRT) is a medical diagnostic scanning technique utilizing strong magnetic fields and radio waves to create images of the body.

Natufian   Eastern Mediterranean culture that flourished between 13,000 and 9,800 BCE, before settled agriculture. There is evidence these people planted rye but mainly consumed wild cereals and hunted animals for meat.

Neanderthal   Species of humans closely related (by 0.12%) in DNA to Homo sapiens. Theories abound as to why they became extinct relatively soon after Homo sapiens arrived in their habitat.

Neocortex   The "grey matter" in a human brain, involved in higher functions such as sensory perception, motor commands, spatial reasoning, thought and language.

Neurodiversity   Concept that suggests that neurological differences are normal variations within the human genome, including Dyspraxia, Dyslexia, Attention Deficit Hyperactivity Disorder, Dyscalculia, Autistic Spectrum, Tourette syndrome and other conditions commonly viewed as inherently pathological.

Noosphere   Concept used by Vernadsky and Teilhard de Chardin to denote the "sphere of human thought," a phase of development following the geosphere and biosphere. As life has transformed the geosphere, so has thought transformed the biosphere.

Ordovician   Geologic era, the second of six within the Paleozoic, occurring between 485.4 and 443.4 million years ago, named for the Ordovices, a Celtic tribe.

Paleoanthropologist   Scientist who combines disciplines of paleontology and physical anthropology to study ancient humans from fossil hominid remains.

Photosynthesis   Process used by plants to convert light energy, normally from the sun, into chemical energy to fuel the organisms' activities.

Picoplankton   Found in all oceans and lakes, smaller than other plankton, they have a greater surface to volume ratio enabling them to obtain scarce nutrients in ecosystems known as oligotrophic gyres.

Pirahã   Indigenous tribe of hunter-gatherers in Brazil's Amazonia, mainly living on the banks of the Maici River.

Pleistocene   Geologic era, lasting from 2.6 million years ago to 11,700 years ago, spanning the world's recent repeated periods of glaciations and Ice Ages.

Pliocene   Geologic era preceding the Pleistocene, extending from 5.3 to 2.58 million years ago.

Promethean   Adjective denoting extraordinary size and strength, derived from Prometheus, a Titan engaged in a cosmological struggle with the Olympian Gods of Ancient Greece.

Purgatorius   Genus of the four extinct species believed to be the earliest example of a primate, dating back as far as 66 million years ago.

Quantum Computers   Devices that make direct use of quantum-mechanical phenomena to perform calculations and can operate in more than one mode simultaneously, unlike digital computers.

Radioactive fallout   A highly dangerous form of radioactive contamination, residual radioactive material propelled into the atmosphere following a nuclear blast, so called because it "falls out" of the sky.

Radioactive isotope   Any of several variants of the same chemical element having a different mass, whose nuclei are unstable and dissipate excess energy by spontaneously emitting radiation in the form of alpha, beta, or gamma rays.

Radionuclides   Synonym for radioactive isotopes (see above).

Sahelanthropus   Extinct hominine species dated to about 7 million years ago, close to the time of the chimpanzee/human divergence. Few fossil specimens have been found.

Silk Road   Ancient trade route, 4,000 miles long, between Europe and Asia, significant for the political and economic development of China, India, Persia, Europe and Arabia.

Social network   Online community of people who use dedicated websites or other technologies to share information, resources, images and the like.

Stratigraphers   Geologists who study rock layers (strata) and layering (stratification).

Stromatolites   Layered mounds, columns and sheet-like sedimentary rocks, formed by the growth of layer upon layer of cyanobacteria, in shallow water, They provide the most ancient records of life on earth with fossil remains dating back more than 3.5 billion years ago.

Sustainability Term used to describe a way for humans to live in harmony with the carrying capacity of the earth's ecosystems.

Technosphere Collective term for all aspects of the physical environment that have been created or significantly altered by humans.

Tectonic plates Individual sections of the earth's crust and uppermost mantle.

Telluric force Extremely low frequency electric current that moves underground or through the oceans, over vast areas, at or near the surface of the earth.

Terawatts Measurement of electrical power equaling one trillion watts. The total power used by humans worldwide, in 2006, was about 16 terawatts

Thermodynamic Balance State of equilibrium or balance where there are no net inflows or outflows of matter or energy within a system.

Treptichnus Pedum Earliest widespread complex trace fossil, used to delineate the change from the Ediacaran to the Cambrian geological era.

Tsunami Japanese term for a massive ocean wave caused by displacement of a large volume of water, generated by an earthquake, volcanic eruption, underwater explosion or other disturbance.

UNCTAD United Nations Conference on Trade and Development, a body of 194 states, responsible for development issues, particularly international trade and investment.

UV Ultraviolet (UV) light is electromagnetic radiation with a wavelength shorter than visible light. Excessive exposure can result in harm to the skin, eyes and immune system of humans.

Warsaw Pact Collective defense treaty of the Soviet Union and its satellite states in Central and Eastern Europe, in existence during the Cold War.

World Wide Web Global system of interlinked hypertext documents containing text, images or other media that can be accessed via the Internet, by means of a web browser.

ABOUT THE AUTHOR

 CHRISTIAN SCHWÄGERL, an award-winning journalist, has written for leading national media in his home country of Germany since 1988, covering politics, science, the environment, economy and cultural affairs. He began writing about global environmental policy in 1995 while covering the first UN climate conference in Berlin, and subsequently worked for several prestigious publications there, reporting on the environment, agricultural development and politics internationally.

Since 2012 Schwägerl has helped initiate two major cultural endeavors around the Anthropocene in Germany: The Anthropocene Project, at the federally funded House of World Cultures in Berlin, and a special exhibition at the German Technology Museum in Munich, serving as an external curator. Schwägerl lives in Berlin, where he loves being out in nature with his two children, practicing yoga, bird-watching, and cycling.

ALSO FROM SYNERGETIC PRESS

ESSAYS ON GEOCHEMISTRY AND THE BIOSPHERE
by Vladimir Vernadsky, Introduction by Alexander Yanshin

WHAT HAS NATURE EVER DONE FOR US
How Money Really Does Grow On Trees
by Tony Juniper, Foreword by HRH The Prince of Wales

THE WASTEWATER GARDENER
Preserving the Planet One Flush At a Time
by Mark Nelson, Foreword by Tony Juniper

ME AND THE BIOSPHERES
A Memoir by the Inventor of Biosphere II
by John Allen

Synergetic Press & Global Printing, Sourcing & Development (Global PSD), in associ-
ation with American Forests and the Global ReLeaf programs, will plant two trees for
each tree used in the manufacturing of this book. **Global ReLeaf** *is an international*
campaign by **American Forests,** *the nation's oldest nonprofit conservation organiza-*
tion and a world leader in planting trees for environmental restoration.